Anonymous

The Transactions of the Entomological Society of New South Wales.

V. 1-2, 1863-73

Anonymous

The Transactions of the Entomological Society of New South Wales.
V. 1-2, 1863-73

ISBN/EAN: 9783337328993

Printed in Europe, USA, Canada, Australia, Japan

Cover: Foto ©berggeist007 / pixelio.de

More available books at **www.hansebooks.com**

THE

TRANSACTIONS

OF THE

𝕰𝖓𝖙𝖔𝖒𝖔𝖑𝖔𝖌𝖎𝖈𝖆𝖑 𝕾𝖔𝖈𝖎𝖊𝖙𝖞

OF

NEW SOUTH WALES.

VOL. II.

SYDNEY:

PRINTED, PUBLISHED, AND SOLD FOR THE SOCIETY

BY J. READING & CO., BRIDGE STREET,

And Sold by TRÜBNER & Co., 60 *Paternoster Row, London.*

1873.

CONTENTS OF VOL. II.

ERRATA.

In pages 58, 60, and 61,—

For Baron de Chandoir, *read* " Baron de Chaudoir."

Description of the Anthicides of Australia, by

REV. R. L. KING, B.A.

[Read 7th January, 1867.]

THE small insects which compose this very natural family are distinguished from other *Heteromera* by the separation of their posterior legs, and by the extension of the anterior edge of the first segment of the abdomen in a (generally) triangular form. This appears the most certain distinction, though even this is shared to some extent by some of the *Pedilides* of La-Cordaire. In this family, however, the intercoxal plate when it exists is small and narrow. Although, therefore, it is separated by some authors, it is united to the Anthicidæ by others. I have met with but one member of the abnormal group, and therefore have not felt it necessary to separate it from the more numerous family to which it is certainly very closely allied.

The Australian species of the Anthicidæ are more numerous than was at first supposed, M. Vereaux having reported that but 9 were found in Oceanic. Instead of 9, no fewer than 48, belonging to 5 different genera, are already known to me as Australian : of all but five I have had specimens under examination. I have been permitted to describe the species in the Museum collection, and in that of Mr. MacLeay. So that together with my own collected at Paramatta, and those received from Gawler in South Australia from my friend Mrs. Kreusler, my list is at least a good instalment, although eventually the number of species may be at least doubled. Mr. Masters collected several species during his late visit to Western and South Australia, while Queensland is represented by several elegant species received by Mr. MacLeay from Port Denison. From Victoria, I have not as yet had any specimens.

Of the Australian species some have a very extensive range, indeed the *Anthicus floralis* is common to Australia and Europe, and several are, to judge by the descriptions, almost identical with species from the Cape of Good Hope. One species *Anthicus bifasciatus* extends from Sydney to King George's

Sound, and would therefore deserve the name *Australis*, had not the name *Australasiæ* been already appropriated to what appears to be a far more local species.

By far the greater number of the Australian species, as indeed of the species from other parts of the world belong to the typical genus *Anthicus*. M. de La Ferte-Senectere in his monograph (which I have followed as closely as possible), has included no less than 208 of this genera, out of 295 of the true *Anthicidæ*. The present paper adds 32 out of 48.

I have not found it necessary to add to the number of genera, though *Anthicus abnormis* might by some have been regarded as entitled to the distinction. The structure of the tarsi in which its greatest peculiarity consists is not always a safe guide to generic affinities—as the late Mr. W. S. MacLeay has well shewn. I have, however, found considerable difficulty in following M. La Ferte-Senectere's divisions and sub-divisions of the genus *Anthicus*. Some of the groups are no doubt very natural, but not others, himself being witness.

Although small in size, yet there are few families which surpass the *Anthicidæ* in the beauty of colour, the elegance of form or the rapidity of motion. I know of no Australian forms except among the Buprestidæ more richly painted than *Anthicus nitidissimus*, none more graceful in its shape than *Mecynotarsus concolor*, more nimble in its movements than *Formicomus agilis*. Their natural habitat too may please the most fastidious. Most of them live in flowers, whence the name of the family ; others are found under wood in grass, especially of a dewy morning ; the remarkable genus *Mecynotarsus* is found under debris in the dried bed of the river Gawler ; and lastly many are captured either flying at sunset, or on the tops of paling fences preparing for flight.

I.—PSEUDO-ANTHICITES.

Genus I.   MACRARTHRIUS.   La Ferte.

MACRATRIA.   Newman.

Sp. 1.   *M. australis.*

Ferrugineus pubescens ; thorace ad medium olivaceo, vix

longiori quam lato, antice rotundato postice sub-contracto et ad basin lateraliter marginato ; elytris sub-parallelis piloso-striatis, ad humeros paulo thoracis basi latioribus, maculis olivaceis elongatis lateralibus obsoletis ; antennis et pedibus castaneis.

Long. .10 poll.

Gawler ; South Australia : " in sand near debris, in the dried river bed." *Mrs. Kreusler.*

The intercoxal plate is long and narrow. The spots on the thorax and elytra are of an olive colour, but are hardly distinct. This species has much the appearance of *Dircœa*, and comes near the Indian *M. concolor*.

II.—Anthicites.

Genus II. Notoxus. Geoffroy.

This genus which has its representatives in many parts of the world is remarkable for the prolongation of the fore part of the thorax in the shape of a horn over the head. A single Australian species has been thus described *by Mr. Hope, from Adelaide.*

Sp. 2. *N. australasiæ.* La Ferte-Senectere.

Nigro-piceus, villosus, elytris profunde et parum confertim punctatis, pone humeros transversim depressis, ibique macula una flava singulari ornatis ; antennis pedibusque sature ferrugineis.

Long. .13 poll.

Adelaide.

I have not seen this species. It belongs to the division which has the apex of the elytra of both sexes oblique truncate. The single macula near the shoulder on each elytron can hardly be mistaken.

Genus III. Mecynotarsus. La Ferte.

This genus, though possessed of the thoracic horn so remarkable in *Notoxus*, has been separated from that genus in consequence of the prolongation of the posterior tarsi. I have not

met with it in this neighbourhood; but I have received three
species from my friend Mrs. Kreusler of Gawler, South Australia. They were all captured under debris in the dried bed of
the river Gawler. Other species might probably be captured in
this colony, if looked for under rubbish near the sea beach.

Sp. 3. *M. Kreusleri.*

Nigro-cinereus; elytris sub-parallelis brunneis fasciis duabus
albis, antennis et pedibus ferrugineis.

Long. .10 poll.

Gawler; *Mrs. Kreusler.*

The fasciæ are broad, and apparently composed of whitish
setæ; at times they are almost separated at the suture, and
appear as large maculæ.

I have named the species after my friend Mrs. Kreusler, who
was, I believe, the first to discover them as well as the two
following species.

Sp. 4. *M. ziczac.*

Cinereus; elytris convexis fasciis nigris anguloso-undulatis
notatis; antennis et pedibus incano-ferrugineis.

Long. .10.

Gawler; South Australia. *Mrs. Kreusler.*

The marking of the elytra differs somewhat in different specimens, but generally there are two small spots behind the shoulder
of darker hue than the rest of the elytra. These spots are sometimes united, while a zigzag fascia at the middle is connected
at the suture with the dark apex.

Sp. 5. *M. concolor.*

Totus cinereo-castaneus, elytris parallelis.

Long. .10 poll.

Gawler; South Australia. *Mrs. Kreusler.*

An exceedingly graceful species. The whole insect is of a
cinereous chestnut colour with a very silky pubescence over thorax
and elytra. There is a faint deepening of colour at the shoulders
and at the middle of the elytra which may possibly in some

varieties amount to spots, but which do not in the numerous specimens sent me by Mrs. Krensler.

Genus IV.   ANTHELEPHILUS.   Hope.

This genus was separated from the succeeding one, *Formicomus*, to receive the species without any apparent angle at the shoulders of the elytra.   M. La Ferte-Senectere appears to desire to re-unite the species to *Formicomus*.   Not having seen any of them, I am not in a position to give an opinion.   Judging however from the plates of the species, I hardly think there is a necessity for the division.   A single Australian species has been described by Hope in the following terms.

Sp. 6.   *A. cyaneus.*   Hope.

Capite nigro, antennis pedibusque atris.

Long. 2 lin., lat. ½ lin.

Nova Hollandia.

Antennæ nigræ, articulo basali crasso, reliquis extrorsum crassioribus; thorax ovalis, antice posticeque contractus nigro-cyaneus; elytra cyanea, nitida, glaberrima; corpus subtus nigrum, pedes concolores.

Trans. Zool. Soc. Vol. I., page 101; Plate XIV., fig. 4.

Genus V.   FORMICOMUS.   De la Ferte.

FORMICOMA.   Motzch.

This genus has been established for the reception of the *Anthicites*, which have at the same time the thighs claviform, and the elytra convex and oval.   The intercoxal plate is generally large and obtuse, (truncate or ogee shaped), while in *Anthicus*, it is generally triangular and acute.   The species composing this genus are among the largest of the whole family, often very nimble and sometimes exceedingly handsome.

Sp. 7.   *F. Clarkii.*

Nitidus parce pilosus, capite nigro polito antennis piceis; thorace rufo-flavescente; elytris nigro-cyaneis post humeros

nonnihil depressis, postice rotundatis ; femoribus nigris ad
basin rufescentibus, tibiis rufescentibus.

Long. .14 poll.

Gawler ; South Australia.   *Mrs. Kreusler.*

This species comes very near *F. rubricollis Dej.* from South
Africa, but differs in colour of feet and puncturation of head.   The
last joint of the antennæ is sometimes rufous.   The thorax (at
least in dead specimens) is sometimes piceous at the sides ; it
is rounded in front but regularly contracted towards the base,
yet without the sudden contraction so frequently observable.

I have named the species after the Rev. Hamlet Clark, who
has described so many Australian insects.

Sp. 8.   *F. agilis.*

Ferrugineus politus subglaber tenuissime punctulatus ; thorace
     longitudinaliter canaliculato, ad basin valde contracto
     minute tuberculato ; elytris nigris humeris albo-maculatis
     pedibus piceis.

Long. .15.

Paramatta ; Liverpool Plains.

The antennæ are piceous at the base, but gradually becoming
darker towards the extremities.   The longitudinal furrow and
the tubercles at the base of the thorax, together with the oblique
white maculæ on the shoulders of the elytra, distinguish it at a
glance from all its Australian congeners.   Excepting that there
is a slight projection at the shoulders, this species will come very
near *Anthelephilus ruficollis.*   It appears however to be a true
Formicomus.

Sp. 9.   *F. Denisonii.*

Ferrugineus, nitidus, parce pilosus ; antennis piceis basi
     ferrugineo ; thorace irregulariter punctato ; elytris nigro-
     cyaneis pone humeros depressiusculis ; femoribus basi rufa
     posticis non spinosis.

Long. .15.

Port Denison ; collection of *Wm. MacLeay, Esq.*

A very handsome species from Port Denison. It comes very near to the description of *F. rubricollis* from the Cape of Good Hope, but differs *inter alia* in its ferruginous head. I have named it after the locality in which it was captured ; and am not sorry to have so good a reason for connecting it with the late Governor General of N.S.W., who took so lively an interest in all branches of Natural History.

Sp. 10. F. speciosus.

Piceo-ferrugineus, creberrime punctatus, pubescens setis longioribus paucis, capite nigro; thorace antice rotundato
longitrorsum canaliculato ; elytris convexis nigro-brunneis
quatuor maculis elongatis transversis notatis.

Long. .18.
Gawler ; South Australia. *Mrs. J. Kreusler.*

This fine species appears to approach to the description of *F. aulicus*, from India. It is, however, readily distinguished from that species by the longitudinal division of the thorax, in which it approaches *F. agilis* and *F. canaliculatus*. The maculæ on the elytra are marked by silvery setæ. A single specimen in Mrs. Kreusler's collection kindly lent me by her for description.

Sp. 11. F. quadrimaculatus.

Ferrugineus, subnitidus, pubescens, antennis gracilibus; thorace
antice rotundato postice contracto elytris maculis quatuor
transversis albidis, femoribus parum incrassatis.

Long. .15.
Gawler ; South Australia ; collected by *Mrs. Kreusler*, and by
Mr. Masters.

This species also comes very near *F. aulicus.* It has not the marks on the head peculiar to that species ; and the antennæ are more slender than the description of those organs of the Indian species. The spots on the elytra are transverse rather than oblique. It is easily distinguished from *F. speciosus* by its smaller size, and its thorax not canaliculate. It appears to be common.

Sp. 12. *F. australis.*

Pubescens niger antennis et tarsis ferrugineis ; capite minutissime punctato, postice quadrato ; thorace trapezoidali
superne planiusculo antice transversim rotundato postice
contracto ; elytris humeris subquadratis apicibus rotundatis,
lateribus leviter convexis, maculis albis duabus humeralibus
transversis, duabus pone medium, abdomine elongato.

Long. .09.

N.S.W. Gawler. King George's Sound.

This species, though subject to some variation in the size of
the spots, and the size of the whole insect, is easily recognizable,
and occurs apparently over a large portion of the continent. It
is common at Paramatta. Mrs Kreusler and Mr. Masters have
both found it in South Australia, and there are also specimens in
the collection in the Museum brought by Mr. Masters from King
George's Sound. It has therefore some claim to the name
Australis, by which I have distinguished it.

It is a somewhat aberrant species of this genus, but the
intercoxal plate is large and somewhat ogee shaped. In external
appearance it much resembles the genus *Anthicus*. It is easily
distinguished from the next species by its transverse rather
than oblique spots.

It is found under stones and wood in grass, and is very
active.

Sp. 13. *F. obliqui-fasciatus.*

Pubescens niger antennis et tarsis piceis ; capite punctato
postice rotundato ; thorace antice rotundato postice coarctato ; elytris post humeros depressis, lateribus parallelis,
maculis duabus testaceis ad medium obliquis et albis setis
notatis.

Long. 0.12.

Paramatta ; a rare species.

The thorax is fringed with silvery setæ so as to give it a
somewhat cordiform appearance. At first sight it bears a considerable resemblance to the preceding species. I have not been
able to examine the intercoxal plate. But I place it in this

genus because of its general resemblance to the preceding species. La Ferte appears to lay but little stress upon the shape of the intercoxal plate ; whereas La Cordaire makes it of considerable importance. These two species appear designed to unite *Formicomus* and *Anthicus.* The large intercoxal plate (at least of *Australis*) placing them in the former genus of La Cordaire, and the rather square shoulders bringing them into the *Anthicus* of La Ferte-Senectere.

Sp. 14. *F. senex.* De La Ferte.

Totus nigro-fuscus, subtiliter punctatus, griseo-pilosus ; thorace antice transversim globoso; elytris pone humeros nounihil depressis.

Long. .12.

Nova Hollandia (baie des chiens-marins.)

I have not seen this species in any collection, and simply copy the description from La Ferte. The head is transverse ; and head, thorax, and elytra are all of a fuliginous black, the latter apparently without any spots. It cannot therefore be mistaken for the next species.

Sp. 15. *F. Mastersii.*

Nigro-piceus, creberrime punctatus griseo-pilosus ; capite rotundato ; thorace antice transversim globoso, postice contracto ; elytris maculis duabus obliquis elongatis rufis pone humeros, alteris duabus griseis pone medium minoribus ; antennis et tibiis ferrugineis.

Long. .17.

South Australia; collected by *Mr. Masters*, Australian Museum.

This species evidently comes near the preceding. The anterior maculæ do not appear to be constant. The roundness of the head as well as the size of the insect, and the maculæ on the elytra, readily distinguish it from *F. senex.*

I have ventured to name this fine species after its discoverer.

Genus VI. Tomoderus. De La Ferte.

This small genus has been formed for the reception of the

species of *anthicus* having a short robust form, moniliform antennæ, and a distinctly bilobed thorax. Erichson has marked one from Tasmania, (which La Ferte considers, however, to be identical or nearly so with a European form *T. compressicollis*.)

Sp. 16.   *T. vinctus*.   Erichs.

Pubescens, rufo testaceus ; thorace latiusculo cordato ; elytris striato-punctatis, fascia pone medium nigra.

Long 1¼ lin.

Tasmania.

The original insect is in the Berlin Museum. I have not seen the species. The thorax is transverse, lightly canaliculate on the back, and impressed at the base ; the feet flavous. The black fascia on the elytra alone appears to separate it from *T. compressicollis*.

Genus VII.   ANTHICUS.   Payk.

The genus *Anthicus* is by far the largest of the whole family, at least in point of number. It includes all those which have the head free, the antennæ inserted on the lateral edges of the head, the thorax rounded in front, and more or less strongly contracted posteriorly, but seldom bilobed ; the thighs only moderately if at all enlarged, and the intercoxal plate triangular. Most of the species are subject to very great variations in the size and shape of the spots on the elytra, so that the shape of spots or of fasciæ on the elytra, or even their presence or absence, can never be absolutely depended upon. The colour of the ˙thorax and of the antennæ and feet, and of the ground-work of the elytra is, however, generally constant.

La Ferte has, for the sake of convenient arrangement, divided the whole genus into four great sections, distinguished by the degree of contraction and convexity of the thorax. These are again subdivided into eighteen separate groups, some of which are not represented in this country. The arrangement is probably a merely artificial one ; yet it has its conveniences, and is therefore a considerable assistance in the study of the genus. I shall therefore follow it as closely as possible, although I feel considerable hesitation as to the particular group under which some species should be placed.

La Ferte's first division of the genus is as follows:—

Div.

A.—Thorax without lateral fossa

    a   „   strongly contracted posteriorly   ...   ...   I.

    b   „   slightly contracted posteriorly

          a' disk of thorax, flat and transversely

                 rounded anteriorly   ...   .   II.

          b' convex disk and regularly globular

                 anteriorly   ...   . III.

B.—Thorax with lateral fossa   ...   IV.

Division I.

Group I.

Elytra parallel, hardly convex, elongated, very square anteriorly. Thighs somewhat dilated. Thorax long and bilobed.

Sp. 17. *A. nitidissimus.*

Nitidus, subglaber; capite piceo; thorace ferrugineo, antice rotundato postice valde coarctato; elytris antice punctatis ad humeros quadratis piceis, fascia pone humeros flava et maculis post medium duabus flavescentibus, antennis et pedibus flavis.

Long. .09.

Gawler; South Australia.

A very elegant species, which I place with some hesitation in La Ferte's first group. The elytra are not so much elongated as in the other species ranged in this group—to judge by the descriptions. The thorax is somewhat suddenly contracted at the middle, and swells out again towards the base. The humeral fasciæ of the elytra are somewhat oblique, in consequence of the triangular piceous mark at the scutellum. It is of a more ferrugineous yellow than the maculæ. The species seems to approach *A. Rodriguii.* Latr.

Group IV.

Thorax anteriorly transversely-globose, contracted towards the base, the posterior sides slightly inflated ; the base generally very minutely tubercled.

To this group several Australian species are referable.

Sp. 18. *A. pulcher.*

Ferrugineus subtiliter punctatus depressus, antennis flavis
articulis 7—10 moniliformibus ; elytris piceis fasciis duabus
latis flavis notatis, apicibus rotundatis ; pedibus flavis.

Long. .09.

Gawler ; South Australia.

The fasciæ occupy the greater part of the surface of the elytra, and are often separated at the suture. The whole insect is very depressed. The thoracic tubercles are very distinct. The species appears to approach *A. debilis* from Egypt, and belongs to that division of the group, with elongated sub-parallel and flattened elytra.

Sp. 19.   *A. bombidiodes.* La Ferte.

Anthicus strictus, Erichsen.

Niger nitidus glaber ; thorace pone medium valde compresso ;
elytris maculis duabus altera humerali altera pone medium
minuta, flavo-testaceis, antennis pedibusque fuscis, tarsis
pallidis.   ●

Long. .18 ; (? .08. an error here in the description of *La Ferte.*?)

Adelaide ; South Australia.

I have received from Mrs. Kreusler, from Gawler, two specimens which agree with the above description, except in regard to the size of the insect and the colour of the elytra. I hesitate to regard them as undescribed. The elytra on my specimens are very convex and oval.

Sp. 20.   *A. comptus.* La Ferte.

Piceus, nitidus, capite nigro ; thorace basi ferrugineo ibique
subtilissime tuberculato ; elytrorum fasciis duabus, antennarum basi tarsisque flavis.

Long.

Adelaide and Gawler ; South Australia.

An insect very variable in colour of elytra. Mrs. Kreusler has sent me a specimen from Gawler, and Mr. MacLeay has also received it from Mr. Odewahn, of the same place.

Sp. 21.　*A. unifasciatus.*

Piceo-ferrugineus nitidus thorace bisinuato antice polito postice
　alte punctato ad basin bi-tuberculato ; elytris ovatis punctu-
　latis brunneis, fascia flava pone humeros unica ; antennis et
　pedibus ferrugineis.

Long. 0.10.

Gawler ; South Australia.

A very pretty species near *A. prædator*, distinguished from its Australian congeners, by its single broad yellowish fascia behind the omoplates. There is no appearance of spots elsewhere. It appears to be not uncommon.

The name *unifasciatus* has been used by Schmidt, but as it appears to have been applied to the previously described *A. venustus* (Villa) ; this name so descriptive of the present species is again at liberty.

Sp. 22.　*A. bellus.*

Piceus politus capite nigro subtilissime punctato ; thorace
　pone medium valde compresso ad basin punctato 2-tuber-
　culato ; elytris nigro-piceis lateribus convexis maculis
　duabus humeralibus testaceis, duabus alteris pone medium ;
　tarsis flavis.

Long. .10.

Paramatta ; not uncommon.

It is very near *A. comptus*, but the spots on the elytra, though very variable in size, never appear to unite as in that species. I frequently capture it upon the roof of the house and among the dead leaves and sand in the gutters. It is a very pretty and active species.

Sp. 23.   *A myrteus.*

Pubescens politus punctatus capite nigro ; thorace piceo aut
    ferrugineo, bisinuato, ad basin plerumque bituberculato ;
    elytris piceo-brunneis bifasciatis lateribus sub-convexis ;
    femoribus piceis ad basin ferrugineis.

Long. .10.

Paramatta ; Gawler.

This species appears to have a wide range.   Specimens
which I have received from Mrs. Kreusler, captured in South
Australia, are not to be distinguished from Paramatta specimens.
It frequents flowers, and is very common among the *Leptospermun*
and *Bursaria* when in bloom.   I also frequently find it on roses.

The antennæ are rufous at the base, but piceous at the four
last joints.   The base of the thorax is sometimes so deeply
punctured as to appear almost foveolate.   The fasciæ on the
elytra sometimes cover the greater portion of the surface.

Sp. 24.   *A. glaber.*

Totus castaneus glaberrimus minutissime punctatus capite
    rotundato ; antennis subclavatis : thorace vix ad basin
    tuberculato ; elytrorum marginibus exterioribus piceis.

Long. .08.

Gawler ; South Australia.

This species belongs to the same division with *A. comptus,*
from which it is distinguished by the roundness of its head as
well as its colour.   I can discover no trace of setæ or pubescence
with a strong lens.   The antennæ are more clavate than usual ;
the last three joints gradually increasing in size.

Sp. 25.   *A. intricatus.*

Ferrugineus glaber ; thorace antice transverse rotundato
    postice contracto intricato ; elytris brevibus lateribus con-
    vexis punctatis flavo-ferrugineis et medium et ad apicem
    piceomaculatis lineis suturalibus notatis.

Long. .08.

King George's Sound ; *Mr. Masters,* Australian Museum.

The only specimen collected by Mr. Masters is in the Museum. The intricate markings of the thorax and the short glabrous elytra, distinguish this species very readily from its congeners. The base of the thorax is distinctly tuberculate.

Sp. 26. *A. Denisonii.*

Capite piceo punctato polito antennis ferrugineis articulis ultimis piceis; thorace ferrugineo antice transversim rotundato postice contracto lateribus sub-parallelis ad basin punctato non tuberculato; elytris pubescentibus vix punctatis brunneis fasciis flavis latis duobus, lateribus convexis.

Long. 0.10.

Port Denison; in *Mr. MacLeay's* collection.

Although the thorax is not, or very indistinctly tubercled at the base, yet the general appearance of this species is that of the group under which I have placed it. The latter half of the thorax is nearly cylindrical. The sutural lines are very close to the suture. Another specimen from Port Denison in the same collection, is much lighter in its colour, but in the paucity of specimens, I must place it under the same name.

I have dedicated the species to our late Governor-General, whose name has been given to the locality from which these specimens have been received.

Sp. 27. *A. dubius.*

Pallide ferrugineus politus; thorace binodoso non tuberculato; elytris subconvexis flavo-castaneis pubescentibus; antennarum basi tarsique flavis.

Long. 0.07.

Paramatta; in grass.

I have placed the species in this group very doubtfully. There are no traces of the thoracic tubercles, though the general contour is that of this group. There are hardly any traces of marking on the elytra, except in some specimens a slight deepening of colour principally due to the transparency of these organs, permitting the abdomen to be seen.

Group V.

Thorax rugose, globose anteriorly—the sides generally spiny ; elytra more or less striato-punctate.

Sp. 28.   *A. scydmænoides.*

Castaneus punctatus setosus ; capite piceo antennis castaneis articulis 9 et 10 nigris ; thorace lateribus subangulosis subspinosis ; elytris convexis humeris piceis et duabus pone medium maculis transversis ; pedibus castaneis.

Long. 0.10.

Collection of *W. MacLeay, Esq.*

I believe this species to be near the 5th group; the strong black setæ on the thorax have the appearance of spines ; similar strong black setæ cover the pubescent elytra.   The punctures on the elytra are strong, but not at all arranged in striæ.   The very convex elytra separate it very decidedly from its Australian congeners.

Division II.
Group VII.

Thorax cordiform, transverse ; whole body flavo-testaceous. A small group ; but represented in Australia by certainly two if not three species.

Sp. 29.   *A. luridus.*

Pallide ferrugineus, punctatus, sericeo-pubescens ; elytris luridis immaculatis.

Long. 0.10.

Port Denison.   Collection of *Mr. MacLeay.*

This species comes very near the European form *A. bimaculatus*, but it wants the maculæ on the elytra—as indeed is the case with some specimens of *bimaculatus*.   The head and thorax are decidedly ferrugineous.

Sp. 30.   *A. apicalis.*

Luridus, parce setosus ; capite, antennis et elytrorum humeris et apicibus murinis ; ad medium elytrorum maculis duabus murinis.

Long. 0.10

Port Denison ; *Mr. MacLeay's* collection.

This species also comes very near *bimaculatus* and *luridus.* The elytra are less closely punctate, and therefore I am justified in placing more value than usual on the colour of the elytra in separating the species.

Sp. 31. *A. immaculatus.*

Pallide ferrugineus punctulatus capite transverso postice quadrato ; thorace subtransverso, antice transverse rotundato
postice contracto ; elytris brunneis immaculatis lateribus
rotundatis, lineis suturalibus notatis.

Long. 0.12.

South Australia ; *Mr. Masters.*

The elytra are much rounded, the sutural lines very distinct. There is a faint ferrugineous tint on the shoulders and along the suture. A paler variety I have received from Mrs. Kreusler, captured at Gawler, but it seems to be the same species. There are several specimens of the typical form in the Museum collection, brought by Mr. Masters from his late Western excursion.

Group VIII.

Thorax oblong trapezoidal flattened ; body shining, almost glabrous ; elytra parallel, hardly convex, anteriorly quadrate.

Sp. 32. *A. floralis.* Payk.

Fusco-brunneus nitidus glabriusculus, subtiliter punctulatus ;
antennis, pedibus, thorace, elytrisque antice ferrugineis.

Long. 0.13.

South Australia ; *Mrs. Kreusler.*

This species, for I cannot distinguish any real difference between the South Australian and the European specimens, has not only a large range, but has also been long known. It was described by the great naturalist in his *Fauna Suecica* as *Meloe Floralis* in the year 1735 ; and under different names has been frequently described since. It has been found in Lapland, Scotland, Spain, Turkey; in Algeria, Egypt, and the Cape of Good Hope; in the United States, Chili, and Guadeloupe. Its capture in South Australia, by my friend Mrs. Kreusler, shows that it has travelled even further than had been supposed by La Ferte.

Sp. 33. *A. hesperi.*

Setosus; capite nigro punctulato, occipite bilobato; thorace ferrugineo minute punctato; elytris punctatis lateribus subsinuatis, maculis lateralibus quatuor, duabus pone humeros duabus pone medium obliquis.

Long. 0.12—0.13.

Paramatta; Gawler, South Australia; common in grass; on fences at evening.

The spots vary in size, at times nearly meeting at the suture, at times nearly obsolete. The elytra are slightly convex at the sides, widest near the posterior spots. They are depressed behind the shoulders. The bilobation of the head is sometimes hardly observable.

I cannot distinguish some of the Gawler specimens from those from Paramatta.

Sp. 34. *A. monilis.*

Pallide ferrugineus, subtiliter punctatus pubescens; antennarum articulis 7—10 moniliformibus; elytris piceis, humeris ferrugineis, maculis duabus pone medium flavo-ferrugineis, apicibus rotundatis.

Long. 0.09.

Gawler; *Mrs. Kreusler*, Australian Museum.

The head is somewhat piceous. The moniliform clava of the antennæ and the broad pale ferrugineous fascia on the shoulders of the elytra distinguish it from its congeners. I have received a single specimen from my friend Mrs. Kreusler. A second specimen is in the Museum, collected by Mr. Masters in South Australia.

Sp. 35. *A. Kreusleri.*

Piceus nitidus subtiliter punctulatus setosus; capite nigro

occipite bilobato ; thorace ferrugineo ; elytris lateribus sub-parallelis maculis duabus rotundatis flavo-ferrugineis pone humeros, duabus pone medium ; antennis et pedibus flavis.
Long. 0.09.
Gawler ; *Mrs. Kreusler*, Australian Museum.

In this very pretty species, the shoulders are hardly so square as in the type of the group, the whole insect being narrow. The spots on the elytra are generally round, but at times they are so much enlarged as almost to form fasciæ. The abdomen is not wholly covered by the elytra. I have dedicated this apparently common and pretty species to my friend the discoverer.

Sp. 36. A. charon.

Niger nitidus punctulatus subglaber ; thorace subtrapezoidali antice nigro, ad basin flavo-rufescenti ; elytris immaculatis antice punctatis ; antennis et pedibus ferrugineis.
Long. 0.10.
King George's Sound ; in the collection of *W. MacLeay, Esq.*

This interesting species is evidently near *A. infernus*, from Mexico, and *A. stygius* from the Cape of Good Hope. The basal half of the thorax is yellowish red, the rest is black. The anterior portion of the elytra is deeply punctured ; towards the apex the punctures almost disappear. The elytra are broadest behind the middle. The suture is slightly raised posteriorly. The whole insect is very thinly covered with whitish setæ.

Group X.

Thorax subcordate, the length either equal to, or a little greater than the breadth ; elytra varying in colour and puncturation.

Sp. 37. A. crassipes. La Ferte.

Fusco-brunneus, subnitidus, parum crebre punctatus, sub-hirsuto-pubescens ; antennarum basi tibiis tarsisque rufes-centibus elytris maculis duabus obliquis altera pone humerum altera pone medium flavo-ferrugineis ; tibiis maris insolite incrassatis introrsum emarginatis.

Long. 0.10.

Nova Hollandia.

This was one of the species brought to Paris by M. Verreaux. I have not yet been able to distinguish it among the many specimens in my collection. The male is distinguished by the shape of the posterior tibiæ, which are robust, broad, somewhat bent, but not armed with a spine or tooth, as in the next species.

Sp. 38.   *A. brevicollis.*

Setosus; capite nigro transverso polito, antennis ferrugineis articulis ultimis piceis; thorace brevi transversali cordato, antice rotundato postice contracto, minutissime punctato; linea basali transversa impressa; elytris fusco-diaphanis ad latera et apicem, interdum piceo-notatis; femoribus piceis tibiis flavo ferrugineis; tibiis posterioribus maris introrsum spinigeris.

Long. .12.

Randwick.

The markings of the elytra are subject to considerable variation. The male is always darker than the female.

A species from Gawler, sent me by Mrs. Kreusler, agrees in every respect with this description, except that I have not been able to discover the spine on the posterior tibiæ of the male. Many of the specimens are smaller, and some are even larger than the Randwick specimens.

The Randwick specimens were captured in some abundance on the leaves of *Eucalyptus marginata* running about with great activity.   Others were found in the flowers of a *Leptospermum.*

Sp. 39.   *A. glabricollis.*

Ferrugineo-piceus, tenue pubescens; capite nigro antennis ad basin ferrugineis, clavatis; thorace ferrugineo cordato glabro; elytris subtiliter punctatis, duabus fasciis latis flavo-testaceis; femoribus piceis.

Long. .10.

Gawler; *Mrs. Kreusler.*

The thorax is widest close to the head, and then contracts very gradually to about $\frac{2}{3}$ of its length; the remaining third being nearly cylindrical. The fasciæ of the elytra are broad and almost united at the suture. The sides of the elytra are somewhat rounded.

This species is near the preceding, but the thorax is longer.

Sp. 40. *A. crassus.*

Fusco-brunneus, parum crebre punctatus hirsute-pubescens; antennis ferrugineis; thorace transverso obcordato, ad basin ferrugineo; elytris maculis duabus flavis, posterioribus obliquis.

Long. .14.

Gawler; *Mrs. Kreusler.*

This description applies to one of two specimens, the head and thorax of which is blackish brown, and the maculæ on the shoulders are somewhat large. In the second specimen the head and thorax are ferrugineous, and the flavous spots cover nearly the whole elytra, and are widely connected at the suture. It is one of the many species which we owe to the industry of Mrs. Kreusler of Gawler. It comes near the description of *A. crassipes.*

Division III.
Group XI.

Thorax elongated, elytra cylindrico-elongated.

Sp. 41. *A. nigricollis.*

Parce setosus punctulatus niger; thorace antice rotundato bisinuato; elytris maculis flavescentibus 2^{bus} pone humeros et 2^{bus} pone medium, pedibus ferrugineis.

Long. .10.

Gawler; *Mrs. Kreusler.*

The elytra are elongated; the suture is somewhat prominent; the spots behind the shoulder are sub-triangular, those behind the middle almost unite to form a fascia.

The species belongs to the first division of this group, having the sides of the thorax bisinuate.

Sp. 42. *A. Wollastonii.*

Brunneus opacus punctatus sparse pubescens, occipite quadrato
plerumque bilobato ; thorace simplici lateribus non bisinua-
tis ; elytris ad humeros quadratis ad basin et latera piceis ;
macula media prope apicem picea, tarsis posterioribus sub-
elongatis.

Long. .14.

Gawler ; *Mrs. Kreusler.*

The posterior tarsi are as long as the tibia. The species is
subject to considerable variety of coloration, occasionally it is of
a light straw colour without any trace of spots on the elytra.
The bilobation of the occiput is not constant. The elytra are
deeply and closely punctured ; the thorax less deeply. The
punctures on the head are smaller and more scattered.

I have named the species after my old friend, the author of
" The Insects of Madeira " and other standard Entomological
works.

Sp. 43. *A. rarus.*

Piceus punctatus parce setosus ; thorace subelongato antice
rotundato postice minime · contracto ; elytris elongatis
subparallelis 4-maculatis, 2^bus maculis rotundatis pone
humeros, 2^bus pone medium fasciam formantibus, lineis
suturalibus notatis ; pedibus flavescentibus.

Long. .10.

Paramatta.

This handsome species of which I have but a single specimen,
comes very near *A. Wollastonii,* but may at once be distinguished
by its deeper punctures on the elytra and its sutural lines. Like
the preceding species, the punctures on the thorax are minute.

Sp. 44. *A. Gawleri.*

Setosus punctatus ; capite piceo postice rotundato occipite bilobato ; thorace bisinuato ferrugineo, superne convexo ; elytris
piceis, antice alte punctatis, flavo-fasciatis maculis flavis
duabus subrotundatis pone medium ; pedibus flavis, antennis
ferrugineis.

Long. .12.

Gawler. *Mrs. Kreusler.*

The punctures on the base of the elytra are much deeper than those on the apex. The flavous fascia is very close to the shoulder. There is hardly any trace of sutural lines on the elytra. The bilobation of the occiput is very distinct.

Division IV.

Sides of thorax more or less foveolate.

Group XVI.

Thorax oblong, slightly dilated at the base, and therefore bisinuate at the sides.

Sp. 45. *A. Krefftii.*

Ferrugineus politus punctatus; capite postice rotundato; thorace brevi oblongo, antice subtruncato postice vix contracto ad basin parce dilatato, 2^{bus} foveolis lateralibus, ad basin linea transversa; elytris magnis humeris quadratis pone humeros depressis, lateribus subparallelis, lineis suturalibus et fascia obsoleta ad medium notatis.

Long. .14.

Paramatta.

I have captured but a single individual of this species. The foveoles on the sides of the thorax are nearly circular, and apparently connected by a line. The sides of the elytra are nearly parallel, very slightly increasing in breadth, to about two-thirds of their length.

I have dedicated the species to the talented Curator of the Australian Museum, the Secretary of our Society.

Sp. 46. *A. MacLeayii.*

Rufus politus minute punctatus setosus; capite postice bilobato; elytris lateribus convexis fusco-nigris, 2^{bus} maculis pone humeros triangularibus rufescentibus, altera unica pone medium, lineis suturalibus distinctis; antennarum articulis ultimis piceis; pedibus piceis, tarsis rufis.

Long. .12.

Illawarra; *W. MacLeay, Esq.*

This species was found by Mr. MacLeay in dead Palms, on the sea beach at Illawarra. It is very distinct from all its Australian congeners. The foveoles on the thorax are more elongated than in the preceding.

I have named the species after its discoverer.

ABNORMAL SPECIES.
Sp. 47. *Anthicus concolor.*

Totus brunneus opacus setosus alte punctatus; thorace elongato antice transverse rotundato postice lateribus parallelis, ad medium longitudinaliter depresso; elytris striato-punctatis, sutura postice subelevata.

Long. .13.

Paramatta; *R. L. King*: under dead log, with ants.

This species, of which I have but a single specimen, (there is also one in *Mr. MacLeay's* cabinet) hardly appears at first sight to belong to the genus; nor does it on examination appear referable to any of La Ferte Senectere's group, though in some respects it approaches VIII. The punctures are very deep, especially on the elytra. The thorax is contracted posteriorly *above*, but the sides are slightly spread towards the base. The antennæ are stout, scarcely clavate, joints obconical.

I prefer for the present placing it at the end of the genus, as abnormal. So also the following.

Sp. 48. *A. abnormis.*

Ferrugineus punctatus nitidus planus subglaber; capite transverso piceo, antennis piceis articulis obconicis; thorace transverso quadrato lateribus sub-parallelis, ad medium longitudinaliter obsolete canaliculatis; elytris lateribus parallelis fascia picea ad medium, apicibus piceis; pedibus piceis tarsis ferrugineis articulo penultimo minutissimo.

Long. 0.10.

Paramatta.

A very abnormal *Anthicus*, the transverse thorax being hardly contracted at all, and the penultimate joint of the tarsi being almost invisible. It is thinly covered with short adpressed setæ.

On the genus *Charagia* of *Walker*,

By A. W. Scott, M.A.

[Read 2nd September, 1867.]

The Catalogue of the *Lepidoptera Heterocera* contained in the British Museum characterizes in page 1569, the genus *Charagia*, to which generic description is appended a Synopsis of the species known to Mr. Walker; the arrangement being apparently founded on the colour and marking of the superior wings, without taking into due consideration the probability of the existence of any of those striking sexual dissimilarities in both of these respects, which this group, particularly in the smaller kinds, so forcibly exhibits.

As this Synopsis, when applied to the species of the genus, enumerated in the catalogue, and to those additional ones in my collection, now to be described, conveys an erroneous principle, I am desirous of contributing further information than I have already given in the *Australian Lepidoptera*, on this beautiful, but little known, portion of the Hepialidean family, and I therefore gladly avail myself of this opportunity to place in a concise manner before the members of this Society the practical knowledge I possess of the transformations, sexual distinctions, and habits of the Australasian Charagiæ, in the hope by so doing that a clearer and more accurate perception of this restricted genus might be attained.

Genus CHARAGIA.[1]

Charagia. *Walker; Brit. Mus. Cat. Lep. Het., p.* 1569: *Scott. Aust. Lep., p.* 3. Hepialus. *Lewin, Doubleday, Boisduval; Stephen's M.S.S.*

Alæ longæ, sat latæ, leviter falcatæ, apice acuminatæ; angulis analibus valde rotundatis.[2] Caput porrectum. Oculi magni, prominuli. Antennæ brevissimæ, aliquantulum moniliformes, leniter ciliatæ. Palpi labiales distincti, porrecti, triarticulati.

---

[1] Generic characters copied from my work on Australian Lepidoptera, p. 3.

[2] Alæ posticæ *non* semi-hyalinæ. Brit. Mus. Cat., p. 1548.

Maxillæ obsoletæ. Abdomen elongatum, alas posticas superans, lateraliter modice compressum, omnibus partibus ejusdem magnitudinis, apice flabellatum. Pedes excalcariti,[1] anteriores magni, validi, tibiis tarsisque dense pilosis ; postici parvi, graciles, tibiis hirsutis, in maribus externe scopatis; tarsi 5-articulatis, fere glabris. Larva carnosa, cylindrica, ad caput incrassata ; capite segmentoque anteriori corneis ; in ligno habitans, plerumque librivora. Pupa lactiflorea, antice squamosa, postice mollis, elongata, annulis serratis.

Wings long, moderately broad, slightly falcate, pointed at the tips and much rounded at the hinder angles. Head projecting. Eyes large and prominent. Antennæ minute, somewhat monoliform, delicately ciliated. Labial palpi distinct, porrected in front, 3-jointed. Maxillæ obsolete. Body elongated, reaching beyond the wings, slightly flattened laterally, nearly of an equal thickness throughout, with the extremity fan-shaped. Legs spurless,[1] anterior and second pairs large and powerful ; tibiæ and tarsi densely pilose; posterior pairs, small, weak, with long hairs on the tibiæ, forming in the males a large brush exteriorly ; tarsi 5-jointed, almost naked. Larva fleshy, elongated, cylindrical, stoutest anteriorly, with head and first segment horny ; living in the interior of trees and subsisting principally upon the bark. Chrysalis yellowish-white, anterior portion squamose, abdominal, soft and elongated with serrated corneous rings.

*Mas....*Alæ anticæ virideo argentes, nonnunquam aureo, fasciatæ.
     Alæ posticæ subcœruleæ.

Foem..Alæ anticæ virides ferrugineo fasciatæ	virescens, *Doubleday.* Scotti, *Ramsay.* lignivora, *Lewin.*
Alæ anticæ virides argenteo fasciatæ	Ramsayi, *Scott.* scripta, *MacLeay.* eximia ? *Scott.*
Alæ anticæ purpureæ viridi variæ	Lewinii, *Stephens.* splendens, *Scott.*
Alæ posticæ rubicundulæ.	

---

[1] Mr. Walker, in the British Museum Catalogue, p. 1569, states that the hind tibiæ have two very minute apical spurs; these have been repeatedly looked for, but hitherto unsuccessfully.

The larvæ excavate to some little depth, cylindrical cells in the interior of the stems or branches of several species of indigenous plants, in which they wholly pass their lives and undergo all the primary changes. Over the external entrance of this habitation they construct a covering, composed of triturated portions of bark and wood held together by silken threads ; the edges of which are closely adherent to the branch, and thus leave no opening for the egress or ingress of the animal. In this particular, the protective covering, although similar in construction and general appearance, differs materially from that formed by the caterpillars of the *Cryptophasæ*, where the lower portions are left unattached to afford to the insects a free passage in order to obtain their natural food, the leaves of the tree within the stems of which they are located. At first, this covering is but a speck, but by the time the larvæ have attained maturity, it assumes an inflated bag-like form of considerable dimensions, and which, in many instances, is so large that the smaller plants are destroyed by the bark having been eaten completely round the main stem.

The chrysalis is placed within this dwelling, head upwards, and, being provided with serrated corneous abdominal rings, and considerable vitality, is capable of locomotion, a power it exercises frequently by moving up and down the walls of its cell with alacrity. At the last metamorphosis, the anterior half of the chrysalis is thrust out of the aperture, when the skin rends asunder, and the perfect insect departs, leaving the exuviæ remaining in that position.

The perfect insects are about the most beautiful of the nocturnal Lepidoptera ; but unfortunately the colours fade quickly after death ; and it is difficult to imagine that the specimens we meet with in cabinets, were ever the beings so brilliant at their births ; consequently no adequate idea can be formed of the beauty of the various species of this group, nor correct descriptions given of the colouring, unless by the examination of very recent specimens. These moths are nocturnal in their habits, and fly with great velocity.

1.—CHARAGIA VIRESCENS.

Charagia virescens. *Brit. Mus. Cat. Lep. Het.*, p. 1569.
Charagia rubroviridans. *Brit. Mus. Cat. Lep. Het.*, p. 1570.
Hepialus rubroviridans. *Stephens M.S S.*
Hepialus virescens. *Doubleday (Duffenbach's New Zealand.)*

Male.........length of wings 49 lines : of body 26 lines[1].

> *Superior wings*, lovely bluish green, relieved by various silver markings, which consist of :—a line along the basal portion of the costa; an irregular broadish obliquely transverse band a little beyond the middle ; another semicircular one nearer to the base, with numerous others, small and faint, disposed transversely between the nervures and terminating in lunules at the exterior border.
>
> *Inferior wings*, delicate bluish-white.
>
> *Abdomen* and *thorax* bluish-green.   *Eyes* reddish-brown. *Antennæ* tawny.

Female......length of wings 62 lines : of body 28 lines.

> *Superior wings*, bright green with numerous distinct irregular reddish-brown bands and lines mostly disposed transversely.   The costa is barred with green and brown, and the whole wing edged by reddish-brown, the exterior portion being indistinctly scalloped.
>
> *Inferior wings*, pale purplish-red.
>
> *Abdomen* purplish red, becoming green towards the extremity. *Thorax* green. *Eyes* and *Antennæ* similar in colour to those of the Male.

Larva ....length at maturity about 37 lines, is throughout of a pale ochreous tint, with the squamose portions much darker, and the head black-brown.

These caterpillars were found in great abundance near the town of Auckland by my friend Mr. Edward Ramsay, of Dobroyde, near Sydney, when on a visit to New Zealand.

They inhabited the limbs of various trees, and on the stem

---

[1] 12 lines to the English inch.

of one of them, the *Melicytus ramiflorus*, or, " *Mahoc*" of the natives, there were no less than thirty habitations of this species "literally" as he remarked "studded with their abodes."

I feel fully assured that the Charagia virescens and the Charagia rubroviridans of the Brit. Mus. Cat., pp. 1569 and 1570, are the male and female of the one species.

2.—CHARAGIA LIGNIVORA.

Charagia lignivora.   *Brit. Mus. Cat. Lep. Het., p.* 1570.   *Scott's Aust. Lep., p.* 5, *pl.* 2.

Hepialus lignivora.   *Lewin. Lep. Ins. New South Wales, pl.* 16.

Male.........Length of wings 24 lines : of body 15 lines.

> *Superior wings*, vivid emerald green, occasionally yellowish, adorned by a continuous rather broad silver band, running from the base along the costa, to about ⅘ths of its length ; then transversely across the wing to the posterior border, from which it proceeds towards the half of the discoidal cell, and again returns to the interior margin near its base, thus forming in its course a somewhat triangular figure of bright silver over the bed of green.
>
> *Inferior wings*, pale bluish, inclining to a greenish hue ; towards the tips are two short indistinct bars, slightly yellowish.
>
> *Head* and *Collar* yellowish white : *tufts* on the thorax emerald-green. *Abdomen*, upper part greenish white, central delicate purple, and terminal emerald green.

Female...... length of wings 31 lines : of body 18 lines.

> *Superior wings* bright light green over which are delicate irregular lines of scarlet, those under the costa and across the wing a little beyond the middle, are much the most distinct. The outer angle and the interior portion of the base of the wing, are both occupied largely by deep purplish-red, each of these patches relieved by spots of lighter and brighter colour placed within them.

Inferior wings pale yellowish-red.

Body yellowish red; the *tufts* on the abdomen, the *collar*, and the *head* being silver-grey.

Larva ....about 24 lines in length, of a dark cream-colour throughout, with the head and squamose portions darker.

The larvæ inhabit the interior of many plants, such as the Casuarina, Callistemon, Eucalyptus, Dodonæa, Acmena, &c., and are plentifully found within a few miles of Sydney, the lower Hunter River district, and many other localities of New South Wales.

Lewin, in his work on the Lepidopterous Insects of New South Wales, has figured two females, representing them as of different sexes. The correct description of the male of his species, the lignivora, is now given.

3.—CHARAGIA LEWINII.

Charagia Lewinii
Charagia Lamberti   } *Brit. Mus. Cat. Lep. Het.,   p.* 1570.

Hepialus Lewinii
Hepialus Lamberti   } *Stephen's M.S.S.*

Male.........length of wings 21 lines; of body 12 lines.

Superior wings, bright emerald green relieved by various silver markings, namely, a line from the base along the costa to about ⅗ths of its length; from this point transversely across the wing to the margin of the inner border, then back towards about half of the discoidal cell, where it nearly meets another short band, proceeding from the basal portion of the interior margin, thus having formed an almost right angle immediately under the discoidal cell: within the space embraced by the two latter bands are placed two small transverse marginal streaks, and likewise over the outer angle a distinct spot is seen. These silver markings are brought out in strong relief by a shading of purplish brown.

The disposition of the silvery lines and colour of the

wings assimilate greatly to those of the *Ch* : *lignivora*, but are infinitely more slender and delicate.

Inferior wings, pale bluish assuming a purplish tinge towards the anterior border.

Female......length of wings 24 lines ; of body 14 lines.

Superior wings, bright rich purple ; a large green band on the middle, deeply notched in front ; dilated and angular behind.

Inferior wings, pale rich purple.

The *body* in colour throughout, similar to the wings, but darker towards the lower extremities.

Larva ....in length about 19 lines ; is of a cream-colour, slightly pinkish in parts ; the head is black-brown, and the squamose portions pale reddish-brown.

The larvæ are common in the vicinity of Sydney, usually occupying the main stems of the small saplings of the casuarinæ.

The Charagia Lamberti of the Brit. Mus. Cat. is the male of this species, for which we have retained the name of *Lewin*, originally bestowed on the female insect by *Mr. Stephens*.

4.—CHARAGIA SPLENDENS.

Charagia splendens.　*Scott, Aust. Lep., p. 6., pl.* 2.

Male.........length of wings 26 lines : of body 15 lines.

Superior wings, bright yellowish green, mottled with darker, and gaily adorned by numerous complicated markings. A continuous band of silver proceeds along the costa to about ⅔rds of its length, crosses the wing a little beyond the middle to its inner margin, thence towards the base in a zig-zag manner, forming in this latter course a couple of distinct angles. Two silver bands, connected at their upper ends by a curve, run parallel to the exterior margin, and between these and the first described transverse band are two others of a bright, light, silvery-bluish green. On the centre of the wing is placed conspicuously a rather large V shaped figure, also of bright bluish-green.

Inferior wings, lustrous bluish white.

Thorax green ; *head, collar,* and *tufts* on thorax silvery. *Abdomen* bluish white, with an oblong green stripe towards the extremity.

Female ......length of wings 33 lines : of body 20 lines.

Superior wings, the centre occupied by a large triangular shaped patch of vivid light satin green, deepening exteriorly, whose lower angle reaches the margin of the inner border, and whose basal portion immediately under the costa, bears three distinct notches. The apical angle to about half of the exterior margin also displays a broad mark of the same intense green possessing a deep indentation on the inner side. With the exception of a distinct spot of green near the outer angle and two others adjoining the base, the remaining portions of the wing are of a deep rich purple.

Inferior wings, pale purplish red, becoming darker towards the hind angle.

Head and *thorax,* reddish brown ; *abdomen,* pale purplish red, deepening towards the extremity.

Larva ......is much larger, but in other respects similar to that of the *Ch : lignivora,* and these two species occupy in common the plants before enumerated.

We may remark that subsequent to the foregoing descriptions, much larger and finer specimens have been obtained.

5.—CHARAGIA RAMSAYI

Male......... length of wings 51 lines : of body 27 lines.

Superior wings, light satiny emerald green, adorned with various large silvery spots, edged around by black-brown, sparingly disposed along the costa and in an oblique transverse row a little beyond the middle. Three small marginal lunules near the outer angle, two small oval spots on the discoidal cell, a dental marking towards the base of interior margin, and between these two latter, a much curved irregular line,—all of silver.

Inferior wings, bluish, slightly yellowish towards the tip.

The *body* pale emerald green with two large spots of silver on the lower part of the thorax ; *tufts* of hair on upper parts of abdomen, silvery ; *eyes* purplish.

Female ......length of wings 66 lines : of body 33 lines.

Superior wings bright grass green, relieved by large, very bright silver spots brought out in strong relief by an edging of black brown, disposed in a similar manner to those on the male ; but being larger and brighter they are more conspicuous and striking. The curved line, before described, becomes here a large spot.

Inferior wings yellowish-red.

Body and *head* similar in colour, but paler, to the superior wings, bearing two reddish-spots on the thorax ; the *tufts* on the abdomen are yellowish-red, and the *fan-shaped* extremity purplish.

Larva . .....length about 42 lines : creamy-white throughout except the segments over the true feet and the head, which are yellowish-brown ; pinkish annular lines, also, between each segment.

The larvæ live within the stems of the acmena, alectryon, and a few other plants, and were by no means uncommon on Ash Island, Hunter River, when I resided there.

6.—CHARAGIA SCRIPTA.

Charagia scripta. *W. MacLeay, M.S.S.*

Male.........length of wings 35 lines : of body 18 lines.

Superior wings, basal moiety emerald green, exterior moiety lustrous yellowish-green, separated from each other by an oblique transverse band of silver, scalloped within ; the whole surface adorned with numerous labyrinthic silvery lines and bands. The inner half is thickly studded over with short lines of silver, principally disposed transversely ; the other by three bands, also of silver, which run parallel to the exterior margin ; the outer one assuming a chain-

like pattern : the exterior marginal border is likewise deeply silvered.

Inferior wings bluish-white.

Thorax and *head* emerald green ; *eyes* purplish ;
abdomen bluish-white, with silvery tufts.

Female ......length of wings 48 lines ; of body 26 lines.

Superior wings bright grass-green with numerous
intricate markings of much lighter colour, principally
on the basal half. Two oblique transverse rows,
beyond the disc, of large bright spots of silver, each
one placed between the veins, with the exception that
in the external row, between the 2nd and 3rd median
nervules no spot exists, and also three or four others
which adjoin the exterior angle, these, together with
a tooth-shaped marking nearer to the base, and the
delicate silver lines across the costa, complete the
ornamentation of the upper wings of this peculiarly
handsome insect.

Inferior wings yellowish-red inclining to pale yellowish-
green towards the tips.

Head, thorax and *abdomen* bright green ; *tufts* of
yellowish-red hairs cover the upper portion of the
abdomen, excepting the three ultimate segments :
the *fan-shaped extremity* is also . furnished with
similar reddish hairs.

Several chrysalids in the wood were brought from King
George's Sound, Western Australia, in 1861, for W. MacLeay,
Esq., of Elizabeth Bay ; and in whose collection the perfect
insects are ; from these specimens the foregoing description has
been taken.

7.—CHARAGIA SCOTTI.   ♀

Charagia Scotti.   ♀   *Ramsay, M.S.S.*

Female ......length of wings 54 lines : of body 26 lines.

Superior wings bright grass-green, delicately dotted
over with purplish-brown spots : a slight purplish-
brown transverse band beyond the middle.

Inferior wings yellowish-red, paler towards the tips.

This insect was captured by Mr. Ramsay at Lismore, Richmond River, and the plumage much injured before it reached me. In this locality Mr. Ramsay found the caterpillars in abundance infesting, among other plants, the nettle tree, (urtica gigas) the native Wistaria, &c., &c., but I regret to say, that the several he had so carefully collected, were all destroyed while in their transit to Sydney.

8.—CHARAGIA EXIMIA. ♂

Male.........length of wings 36 lines : of body 20 lines.

Superior wings bright emerald-green, chastely relieved with numerous markings ; a transverse oblique band of gold a little beyond the middle, but not reaching to either margin ; many short, curved lines of bright silver disposed between the veins ; those to the exterior of the transverse band form a chain-like pattern ; while those to the interior are irregular and labyrinthic.

Inferior wings bluish with a slight shade of green ; ciliæ round the outer angle, golden-brown.

Head pro-thorax and *tippets* similar in colour to the fore wings ; *thorax* and *abdomen* to the hind wings ; *eyes* and *antennæ* dark purplish-brown.

Larva.........length about 42 lines, slightly setigerous, creamy-white with a tinge of purplish-red between the segments.

These larvæ inhabited the small stems and branches of the Dodonæa angustifolia, and were found at Ash Island plentifully. All the larvæ we had collected, excepting the one, were lost, arising from the want of proper and sufficient nutriment, the pieces of wood, in which they were, having become from long keeping hard and sapless. The above measurement was taken from one of the finest caterpillars, with which the rearing proved unsuccessful ; the perfect insect, therefore, whose dimensions are given above, is evidently much undersized, and would probably reach, under favourable circumstances, to between 50 and 55 lines.

The more than usually falcate wings ; the band on the fore wing and the brush of hair on the tibiæ of the posterior leg, being of a golden colour; and the somewhat setigerous larva will readily distinguish this species from any of the foregoing.

Description of a new genus belonging to the family Hepialidæ,
of Stephens,

By A. W. SCOTT, M.A.

[Read 7th October, 1867.]

THE paper on the genus "Charagia" of Walker, which I read
at the last meeting of our Society, was an endeavour to correct
certain important errors existing in the Catalogue of the British
Museum, in relation to this group; to describe, in a concise manner, the habits and metamorphoses of the insects; and to place
on record the existence of several new species. In illustration
of the descriptions I then gave, I exhibited the coloured drawings
of all the species enumerated by me, and I trust I succeeded in
affording a clear perception of a class of insects, so peculiarly
Australasian.

I now purpose to present to the attention of the members a
new and very remarkable example of the Hepialidean family,
and which I hope will prove not only acceptable, but will justify
me in creating a new genus for its reception; one, I think,
readily distinguished by several marked characteristics from any
other, with which I am acquainted.

I may here remark that the fourteen genera which compose
the Hepialidæ, according to the Museum Catalogue, are but
feebly represented by species, no less than five of them having
but one each, and it is therefore fairly presumable that the
family is, at present, but inadequately known.

The magnificent forests and brushes of the temperate and
tropical portions of the globe, thick with underwood, and climb-
ing plants, must prove, if carefully searched, a prolific source for
the production of very many new species of these lignivorous
lepidoptera; and it is to such localities that the intelligent collector will have to look to supply the existing deficiencies.

Before proceeding further, I have to express my regret that
the two examples, male and female, of the insect, now under

consideration, were forwarded to me in an imperfect state, but the numerous component parts, still perfect, render the restoration of the remains, a task of no great difficulty, and a matter of considerable certainty.

The female insect was captured, while at rest on the trunk of a tree, by my friend J. E. Stacy, Esq., while on a journey between Port Macquarie and Newcastle ; to him, therefore, as its discoverer, I dedicate the specific name.

A short time afterwards the male was sent to me in a letter by the late Dr. Stephenson, of Chatham, Manning River, accompanied with the following remarks :—"I found these splendid remains in a spider's web, and as it might be probable you may not have seen the insect before, I have taken the liberty of forwarding them to you."

The Manning River, as you are aware, is a short distance to the southward of Port Macquarie, and in the line of road to Newcastle, so that the two specimens of this rare insect were obtained nearly in the same locality, although at different periods.

HEPIALIDÆ. *Brit. Mus. Cat. Lep. Het., p.* 1548.

Genus. ZELOTYPIA.

Corpus crassum ; abdomen longum, alas posticas superans ; alæ longæ, angustæ, apice sub-acuminatæ, margine exteriore per-obliquo, tantum versus apicem lineis alternis vicibus undulatis ; alæ anticæ ocellatæ, macula discali vitrea.

Fœm. ....caput porrectum : oculi prominuli : maxillæ obsoletæ : palpi breves, tenues : pedes excalcariti ? anteriores validi, pilosi ; posteriores graciles.

Mas. ....Pedes posteriores graciles, valde læsi.

Body thick ; abdomen long, extending beyond the wings ; wings long, narrow, slightly acuminated at the tips, extremely oblique along the exterior border, and crumpled towards the apices ; fore wings with an ocellus on each, whose disc is vitreous.

Female..... Head projecting ; eyes large and prominent ; maxillæ obsolete ; palpi short, slender ; legs spurless ? anterior pairs stout, pilose ; posterior slender.

Male ....Legs, posterior pair, slender, much injured.

The wing veins are similar in structure to those of the genus Hepialus.

ZELOTYPIA Stacyi.

Male . .....length of wings 76 lines ; of body 32 (?) lines[1].

Superior wings dark rich fawn-colour ; on the centre of each, a large dull green-coloured ocellus, encircled by a dark brown line, edged with white outwardly, and bearing within it a sub-diaphanous pearly spot. A broad, irregular, oblique transverse band of silver, crosses the wing slightly beyond the ocellus ; the whole space between this and the exterior margin is occupied by numerous wavy, very fine, distinct, light-coloured lines, disposed labyrinthically over the ground-colour ; the inner portion of the wing is thickly sprinkled over with silver, assuming towards the interior angle a series of ovoidal figures. The costa is very broad and powerful, amply barred transversely with irregular angular bands of silvery-white.

Inferior wings pale bright salmon-colour, darker towards the tips, where they become crumpled.

*Female.. ...*length of wings 117 lines ; of body 47 (?) lines.

Superior wings pale salmon-colour ; the centre of each is occupied by a large, bright, ochreous ocellus, girded by rings of brown and white, and carrying within it a largish sub-diaphanous spot of a pearly hue ; a little to the outward of this latter spot, and nearly in the centre of the ocellus, a black lunule-shaped marking stretches across ; the discal areolet is thickly powdered over with white ; the broad space beyond and below the ocellus, from the tip of the wing to its interior base is fully occupied with numerous chaste wavy lines of reddish-brown, which

---

[1] 12 lines to the English inch.

become clouded towards the tip, but more distinct and vivid towards the basal portion, forming there five conjoined, semi-circular spots of dark brown, relieved by pale-coloured rings. The costa is very large and powerful; dark rich brown with numerous irregular, somewhat angular, transverse bars of yellowish-white.

Inferior wings, salmon coloured throughout. *Head*, *thorax*, and *abdomen*, salmon coloured.

The fanciful generic names usually adopted to distinguish the members of this family, has induced me to apply the equally fanciful one of Zelotypia to this new genus, derived from the male of these monsters possessing large dull-green eyes.

On the " Agrotis vastator," a species of Moth, now infesting
the Sea-board of New South Wales,

By A. W. SCOTT, M.A.

[Read 21st October, 1867.]

AGROTIS VASTATOR.

Agrotis vastator, A. W. Scott, M.S.S. W. B. Clarke, *Sydney Morning Herald.*
Agrotis Spina? Guen: Brit. Mus. Cat., Lep. Het., p. 348, part 10.

Corpus robustum. Proboscis sat longa. Fasciculus frontalis
prominens. Palpi breves, porrecti, pilosi ; articulus 3_{us} longi-
conicus, 2_1 dimidio non longior. Antennæ corporis dimidio paulo
longiores : *Mas.* dimidio apicali pectinatæ apices versus ciliatæ ;
Fœm. simplices graciles. Abdomen alas posticas paulo superans,
planum non cristatum. Pedes anteriores parvi ; posteriores longi,
graciles ; tibiæ anticæ spinosæ ; posticæ calcaribus quatuor longis.
Alæ longæ, sat angustæ, planæ non diflexæ ; anticæ apud costam
rectæ, apice vix angulatæ, margine exteriore sat obliquo.

Body robust. Proboscis rather long. Frontal tuft distinct.
Palpi short, porrect, pilose ; middle joint about double the length
of the terminal and half as long again as the basal. Antennæ
somewhat longer than half of the body ; *Male* pectinated to half
its length, thence ciliated to the tip ; *Female* setaceous, ciliated
beneath. Abdomen extends a little beyond the hind wings,
terminating in a small tuft, flat, not crested. Legs anterior pair
small, tibiæ spinose in front ; 2_{nd} pair longer with two spurs ;
posterior pair long and thin with four long spurs ; tibiæ and tarsi
of all the legs covered with elongated scales and rows of setæ.
Wings long, rather narrow, flat not deflexed when at rest ; fore
wings straight in front, hardly angular at the tips, slightly
oblique along the exterior border.

Length of the body 12 lines : of the wings 24 lines.[1]

Male: fore wings shining light brown, mottled with darker:
a velvety black lanceolate longitudinal discal band, containing

---

[1] A line is equal to $\frac{1}{12}$ inch.

within it, near to each extremity, two spots of light colour bordered by a thin black line, the exterior one reniform and much the largest, the interior one orbicular with, in some, a black centre. Numerous wavy transverse lines, generally very indistinct, but in some specimens well defined: exterior margin entire with a fringe of silvery hue and bordered rather broadly by darkish brown; hind wings light glossy neutral tint becoming towards the outer border much darker, and consequently making the marginal fringe more distinct than in the upper wing.

Female: wings similar in markings to those of the male but the colour is throughout much darker and richer.

The thorax of both sexes is similar in colour to the fore wings, and possesses well defined tippets: the abdomen resembles in hue the hind wings. The sexes can readily be distinguished, by the male being light in colour, and having the antennæ pectinated.

The eggs of lepidopterous insects, when laid in genial weather, will hatch in a few days; those in the autumn will remain quiescent during the winter, and come into existence the following spring. The duration of the Chrysalis stage is likewise extremely variable, and dependent on the difference of temperature. By keeping the egg and the chrysalis in an ice-house, their development may be retarded for two or three years. When removed to a hot-house, ten days or a fortnight will suffice to bring the insect into animated existence: the principle being beautifully illustrated by the late introduction of the ova of the Salmon and Trout into the Colony of Victoria.

The caterpillar of this moth is fleshy, a little attenuated at each extremity, sub-vermiform in appearance, and of a livid colour, varying much in shade, with the anterior segment furnished with a horny plate. It has sixteen feet, measures at maturity about 24 lines, and undergoes its transformation in the ground. The chrysalis is cylindro-conical, of a shining yellowish brown, and protected by a slight cocoon of a rough irregular ovoid form, composed of agglutinated earth.

The genus *Agrotis*[1] even in its limited sense, possesses very

---

[1] Agrotis. Ochsenheimer; Stephens; Boisduval; British Museum Catalogue, Lepidoptera Heterocera 3rd series, p. 303.

many species which have a range nearly world-wide. The caterpillars of several species, such as the one now under consideration,
are very destructive, on account of their numbers, feeding on the
roots and leaves of low herbage, and hiding during the extreme
heat of noon under clods of earth, stones and other convenient
places. Their voracious attacks upon the growing crops in the
field or in the garden, have been for many years past experienced
to a frightful extent, and are too well known to require description.
The number of larvæ in seasons which prove favourable for their
development almost surpasses belief, but our astonishment will
cease when we take into consideration the probable progeny
of those vast numbers of moths now infesting the sea-board of
this land for some hundred miles, each pair producing an off-
spring of many hundreds, who in their turn, before the cold sets
in, will prove equally prolific.   The caterpillar no sooner emerges
from the egg than it begins the great business of life, and falls
vigorously to work—eating—and his growth is marvellously rapid;
"few creatures can equal him in the capacity for doubling his
weight, not even the starved lodging-house slavey when she gets
to her new place, with *carte-blanche* allowance and the key
of the pantry; for in the course of twenty-four hours, he will
have consumed more than twice his own weight of food; and
with such persevering avidity does he ply his pleasant task, that,
it is stated, a caterpillar in the course of one month, has increased
nearly ten thousand times his original weight on leaving the egg;
and to furnish this increase of substance, has consumed the pro-
digious quantity of forty thousand times his weight of food—
truly a ruinous rate of living!"[1]   A few years ago on the Hunter
River, I carefully examined a paddock of twenty-five acres, under
oats for hay, which was much infested by the caterpillars of this
species, and found that nearly every stalk had at least one caterpillar on it, numbers had two; many three : taking the plants at
twenty to the square foot, and each with only one caterpillar, the
result would be 21,780,000 of these insects, and supposing that
all these lived to become moths, each pair producing by the end
of the season a progeny of 80,000,[2] the total produce for the

---

[1] British Butterflies, by W. S. Coleman, 1860.
[2] Reaumur.

twenty-five acres would amount to 871,200,000,000. What then, calculating under the same conditions, would be the number of the caterpillars, which were, at the time I allude to, ravaging whole districts ?—a long line of figures almost unpronounceable.

Allowing for every reasonable loss caused by weather, not unusually severe, accident, or by their numerous enemies, still there would remain quite sufficient to produce those vast numbers of moths, collected together from a wide range of country and seen clustering in caves, under ledges of rocks, in churches, houses, barns, in every nook and cranny where their gregarious habits lead them to, seeking shelter from the glare of day. I, therefore, think that this natural increase, aided by favourable weather, is quite sufficient to account for the swarms of moths recently seen in many localities, and remarked upon by several correspondents of the "*Sydney Morning Herald*" without having recourse to improbable theories. All moths are in their primary stages purely terrestrial, and cannot "come in from the sea" in the sense used by a writer in the "*Newcastle Chronicle.*" They cannot be born there, neither are their wings adapted for so long a flight as to cross the ocean from any point of land to the eastward of our coast, particularly "in the teeth of westerly winds." Indeed many swarms of insects, besides the lepidoptera, are known to be blown from the land, while a few others wilfully fly seaward under some unaccountable, almost insane, desire ; but all these inevitably perish. Mr. Lindley, when at Brazil, in 1803, saw an immense flight of butterflies for several days successively, which were observed never to settle, but flew in a direction from north-west to south-east, direct towards the ocean where they must certainly perish ; and Mr. Barrow in 1797, writes "the locusts covered an area of nearly 2,000 square miles were driven into the sea by a north-west wind, and formed a bank three or four feet high, and when the wind was south-east, the stench was so powerful as to be perceptible at the distance of one hundred and fifty miles."

Without multiplying instances I would suggest that the moths seen by vessels at sea were either endeavouring vainly to emigrate, or, what is *far more probable*, were driven away from the land by the prevalent westerly winds, and perished by

thousands in the ocean : those seen returning to the shore were
the fortunate few that had escaped before being carried too far
to sea. I remember some years ago walking along the sands for
about five miles, between Newcastle and Red Head, and I
observed an almost continuous undulating line of dead bodies,
several deep, of these moths, marking the wash of high water
along the whole of this length of beach, interrupted only by
the rocky headlands, and probably this exhibition of the fate of
these insects in such vast numbers was continued for a considerable distance on either hand.

Were it not for the wholesale destruction of these vast
assemblages of insect pests, caused by the violence of winds—
by the fall of rain for several days successively—by sudden
change of temperature—and by the host of enemies, following in
their wake, consisting of insectivorous birds, and reptiles, and
the numerous family of the Ichneumonidæ, I fear all the
endeavours of man by artificial means to eradicate them would
be baffled. The abundant food furnished by the roots and leaves
of the various weeds and grasses growing over a vast extent of
waste lands, will always ensure too ample a supply of such
noxious creatures. We can, however, check in some degree the
injury to our crops, and thus moderate the evil ; by ploughing
and harrowing the fallow lands, thus cutting off the immediate
supply of food ; by passing the roller again and again over the
growing crops when practicable ; and by encouraging, not
molesting, the many species of birds that visit the fields in flocks
on such occasions. I have seen crows, large brown hawks,
magpies, cranes, spur-winged plovers, and a host of smaller
birds, enjoying during the day ample meals furnished by these
caterpillars, and had a great difficulty in preventing the overseer
from driving them away " because," he said, " they eat the
lucerne." The large family of Ichneumons (little wasp-like
creatures) is also a great ally of man in the war of extermination,
for they pierce the bodies of the living caterpillars, depositing
their eggs within them, and thus cause a slow but certain death
before the larvæ can attain to the perfect or winged state, and
on this account they ought to be encouraged. I add a few words
to assist in that object, although with but faint hope of success.

The cocoons of the Ichneumons are silky, small, oval and yellowish ; attached in groups to walls, palings, and frequently over the remains of dead caterpillars : " these," Mr. Westwood observes, " ignorant people mistake for the eggs of the caterpillar, and destroy, foolishly killing their benefactors."

The present season, dry and warm, has been unusually prolific in the production of these insect pests, whose gregarious habits have been so well described by the Rev. W. B. Clarke of St. Leonard's,* who, *inter alia*, says that the state of St. Thomas' Church, North Shore, on the 14th September, from the enormous numbers of moths, was such that Divine service could not be held therein ; that seven days hard labour in endeavouring to subdue them had been spent in vain ; and that he had counted more than 80,000 grouped together on the windows. Accounts from Newcastle, 70 miles to the north, Wollongong 40 miles to the south of Sydney, and other distant parts confirm this statement as to numbers, and clearly point out *what has been*, and assuredly leads us to the question *what will be ?* should the weather continue as it is. Complaints have already reached us from Windsor, of whole fields of young Lucerne being destroyed by caterpillars, and the farmers appealing to the public for relief. The intelligent agriculturist will accept this warning, take time by the forelock, and quickly adopt such means as may be at his command—for half a loaf is better than none.

The remains of a moth which Mr. Clarke captured in 1851, near the summit of the Mount Kosciusco range in the Australian Alps, was sent to me by that gentleman for comparison with those moths now so abundant around us. I have placed this mutilated specimen under the microscope, and I believe it to be identical with the Agrotis above described. Mr. Clarke assures me that this insect was the species so celebrated for being the food of the aboriginals of that large district for many years gone by, and known by them as the Bougong. I have never visited the Upper Tumut, and know nothing personally about the history of these very remarkable moths.

---

* See Rev. W. B. Clarke's interesting letter in the " Sydney Morning Herald," 11th October, 1867.

I therefore add a few lines descriptive of their habits which I obtained from a source unmistakeably accurate, and which I hope will prove interesting.

In January and March of the year 1865, my friend Mr. Robert Vyner visited the Bougong Mountains, accompanied in the first instance by an aboriginal, " Old Wellington," and in the other by Mr. Sharp of Adelong, Old Wellington, and another black-fellow, both of these latter well acquainted with the habits of the moth, *called* by them " Boogong " and " Gnarliong," indiscriminately. The tops of these mountains are composed of granite, and present a series of lofty peaks, and it was up one of these, named by the natives " Numoiadongo " he and his companions toiled for nearly six hours before attaining the summit; so steep and rugged was the path that even the wild cattle never attempted to ascend to these heights.

The moths were found in vast assemblages sheltered within the deep fissures, and between the huge masses of rocks, which there form recesses, and might almost be considered as " caves." On both sides of the chasms the face of the stone was literally covered with these insects, packed closely side by side, over head and under, presenting a dark surface of a scale-like pattern— each moth, however, was resting firmly by its feet on the rock, and not on the back of others, as in a swarm of bees. So numerous were these moths that six bushels of them could easily have been gathered by the party at this one peak; and so abundant were the remains of the former occupants that a stick was thrust into the *debris* on the floor to a depth of four feet. Mr. Vyner tells me that on this occasion he ate, properly cooked by Old Wellington, about a quart of the moths, and found them exceedingly nice and sweet, with a flavour of walnut, so much so that he desires to have " another feed." His clothes, by the moths dashing against them on being disturbed, were covered with honey, and smelt strongly of it for several days. At the time these multitudes assembled, the tea tree and the small stunted-looking white gums were in full blossom, no doubt yielding up their honied treasures to these nocturnal depredators, whose flight, when issuing from their hiding places to the feeding grounds, was graphically described by Old Wellington " very much like wind, or flock of sheep." The Tumut blacks report that the moths do not congregrate on the high peaks in the spring time, but they first locate the lower mountains, feeding on the blossoms which appear there earlier, and then work their way up to the higher peaks where the plants are later in bloom.

The Bougong moths are collected and prepared for food by the Aborigines, in this wise—a blanket or sheet of bark is spread on the floor ; the moths on being disturbed with a stick, fall down, are gathered up before they have time to crawl or fly away, and thrust into a bag. To cook them a hole is made on a sandy spot, and a smart fire lit on it until the sand is thoroughly heated, when all portions left of the glowing coal are carefully picked out, for fear of scorching the bodies of the insects, (as in such a case, a violent storm would inevitably arise, according to their superstitious notions). The moths are now poured out of the bag, stirred about in the hot ashes for a short time, and then placed upon a sheet of bark until cold. The next process is to sift them carefully in a net, by which action the heads fall through, and thus, the wings and legs having been previously singed off, the bodies are obtained properly prepared. In this state they are generally eaten, but sometimes they are ground into a paste by the use of a smooth stone and hollow piece of bark, and made into cakes.

In this locality were seen many of these holes, having been formed years ago for a similar purpose by the then numerous blacks.

Mr. Vyner also mentions, that at the period of his visit to this peak, he saw hundreds of crows and magpies feeding upon these moths, and the foot marks and other tracks of native dogs and tiger cats were abundant, leading direct to the fissures of the rocks, and although he did not see these animals, he adds, "I am certain from their traces that they must feed upon them," (the moths).

NOTE.—Since the foregoing observations were written, a friend of mine, who resides on the Upper Tumut, forwarded to me a batch of moths, captured on the heights of the Bougong Mountains, purposely, at my solicitation, at the proper season, when the Bougong insect is known to congregate in such multitudes, and when the aborigines in former times were wont to assemble for the annual feast upon their bodies.

I found upon examination of these recently acquired specimens, that they consisted of the males and females of the Oxycanus fuscomaculatus of the Brit. Mus. Cat. Lep. Het., p. 1574; the genus being the 12th of Stephen's family, Hepialidæ. At this result I felt much relieved, for I had made many unsatisfactory enquiries, being doubtful of the genus Agrotis, respecting the habits of the larvæ, which produced in another stage an article of agreeable and nourishing food to the natives of the locality, so plentifully and probably for generations; to these questions, I invariably received for answer, that no assembled multitude of caterpillars, sufficient to account for the vast hordes of Bougong moths, were known in the Tumut district.

It, therefore, appears highly probable that the present insect is the true Bougong Moth; and I give the following reasons for this belief:—the body is plump, very oily and sweet to the taste, characters similarly entertained by most of the species of the Cossidæ and Hepialidæ; and the larvæ, being *under-ground root-feeders*, would not necessarily attract notice, either from their vast numbers or destructive qualities, the latter only exercised upon wild and valueless plants.

These natural conditions are wholly opposed to those possessed by the Agrotis vastator; in whom the abdomen of the perfect insect is neither unctious nor palatable; and whose habits in the larval state are strictly external, and, at uncertain periods of visitation, highly injurious to the interests of man; characteristics perfectly sure to attract the attention of even the most unobservant to their existence.

On the *Ornithoptera Cassandra*,
By A. W. Scott, M.A.

[Read 6th July, 1868.]

My friend, Mr. Edward P. Ramsay of Dobroyde, near Sydney, having sent me for examination a case containing numerous lepidopterous insects, collected for him by Mr. E. Spalding at Rockingham Bay, Northern Australia, during the months of December, January, and February last, I had the gratification of finding in this collection no less than nine males and seven females of the Ornithoptera Cassandra; an insect, the female of which, the male being then unknown, I described and figured in page 131, plate 10, of the first volume of our Transactions, from an individual captured at Port Denison in February 1862, by Mr. George Masters, now assistant Curator of the Australian Museum.

This was the only specimen I possessed to compare with those many nearly-allied species, said to be exclusively confined to Australia, and having a geographical range there from Richmond River, New South Wales, to Cape York, Queensland; a latitudinal extent of nearly 1000 statute miles[1]; and likewise with those inhabiting the adjacent, as well as the more eastern of the Indian Islands, comprehending Woodlark and Darnley Islands, New Guinea, Amboyna, Solomon Islands, &c. As the members of this group closely resemble each other in form and colouring, I, therefore, experienced some difficulty in determining the species to be new: my view in this respect has however now been happily confirmed by the recent acquisition of so many fine examples of both sexes.

Before proceeding to furnish a detailed account of this species, I may be allowed to premise, with respect to the plumage of the female insects, now presented to view by these recently acquired specimens, that I find among themselves, considerable disparity in the size, not in the disposition, of the dull-whitish markings on the anterior wings; and more especially to those previously described; so much so as to necessitate a further description, supplemental to the one already given.

---

[1] The expression " extreme North of Australia," used by Doubleday and Westwood, in defining the range of the Australian Ornithoptera, is incorrect.

D

These very deviations, however, tending as they do uniformly towards the diminution of the spots, as borne by the original specimen, itself comparatively obscure, render the majority of the species still more sombre and consequently more readily distinguishable from others of the family.

ORNITHOPTERA CASSANDRA.

Ornithoptera Cassandra. ♀ *W. MacLeay M.S. Scott Trans. Ent. Soc., N.S. W. Vol.* 1., *p.* 131, *pl.* 10.

Male........length of wings : 74½ lines largest, 67½ lines smallest of the 9 specimens.

Superior wings. Upper surface deep velvety-black, relieved by two broad irregular curved bands of rich satiny green, which spring from the base; the one runs under the costa towards the anterior angle ; the other, along the inner margin and the outer one, as far as the first discoidal nervule : immediately over this latter is placed a large brownish-patch, disposed longitudinally. Under surface, black, with a central spot, and a large macular band, formed of contiguous wedge-shaped spots, placed between the nervules, of gilded green. These wedge-shaped spots are distinctly separated into two divisions by a broad black band. There are, also, two irregular greenish streaks towards the anterior angle, the inner one being short, almost macular.

Inferior wings, upper surface, bright silky-green, with the entire marginal border and four, sometimes, five somewhat large oval spots, disposed between the costal nervure and the first, or second median nervule, one in each space, velvet-black : between these spots and the posterior border are two, or, three minute golden-orange specks, which, however, are not seen in some of the specimens. This tendency to change also exists in the large quadrate golden-coloured space at the immediate basal portion of the anterior margin, shown by some, while in others it is much lessened, or nearly obsolete. Long, fine, closely-set dark-brown hairs spring from immediately under-

neath the inner margin, and rising upwards partially
envelope the upper portion of the abdomen. The
outline of the nervures are easily traceable by narrow,
but distinct, lines of black. Under surface, cor-
responds to the upper, but the green is of a more
golden hue ; the black spots, here seven in number,
become larger and less oval, and the nervures are
broadly picked out with black.

Head and *thorax* deep black, the latter bearing *above*
a central line of satiny-green, and *below* crimson spots
on either side. *Abdomen* bright golden-yellow.

Female......length of wings, 93 lines largest ; 87 lines smallest of
the eight specimens.

Superior wings, upper surface rich black-brown, relieved with various irregular patches and spots of
impure white, similar to, but in lesser degree than
any of its congeners. Three of these, more or less
developed, are placed in the discoidal cell ; one, rarely
two, immediately under ; and a series beyond these,
running obliquely across the wing, exhibits in some
specimens (see plate 10, vol. 1) largish and distinct
markings ; in others, small and faint ; in the first
case they become parted into two at the discoidal
nervules, in the other, indistinct, almost obsolete. A
few small spots, absent in some, along the outer
margin, complete the whole relief of this sombre
insect. Under surface similarly marked to the upper.
Inferior wings, upper surface possesses four wedge-
shaped markings of dusky white placed one in each
space, between the second sub-costal and third median
nervules, although in some specimens, that one
between the second sub-costal and first median
nervules, is obsolete, exhibiting only a small triangular spot at its lower end, the black ground colour
of the wing entirely covering the remaining space.
These markings become dull ochraceous towards their
outer margins, and bear in their centres large, somewhat heart-shaped spots of dark brown, which unite
in the disc with the median nervules.

A sub-quadrate patch of dull ochraceous colour, is placed over the anal angle; and two other spots of brightish yellow are situated between the first and second sub-costal nervules, near to the anterior angle. *Under surface* similar to the upper, but the white is purer, and a bright yellow replaces the ochraceous tint. The margins of the wedge-shaped patches are also entire.

Head and *thorax* dark black-brown, the latter bearing *above* a central longitudinal band of satiny green, and *below* crimson spots on each side. The *abdomen* dark black-brown, grayish towards the tip, and broadly barred underneath with yellow bands.

Habitat: Port Denison and Rockingham Bay, Northern Australia.

In estimating the distinctive characters, I shall limit my comparisons to those species only, which are comprised within Doubleday and Westwood's 1st group of the genus Ornithoptera, and which exhibit the peculiar type of colouring, and occupy the same geographical range of the present insect: purposely excluding the 2nd group, represented by the Amphrisius, Amphimedon, Darsius, Pompeius, and others, from their marked dissimilarity.   •

The Ornithoptera Cassandra differs; from the O. Priamus by being smaller; by the males possessing on the under wings small golden specks over the outer margin, instead of two large ones, and by the absence of one, between the costal nervure and sub-costal nervule; by having on the underneath surface of the upper wings, a distinct double row of gilded-green spots; by the females bearing a green stripe on the thorax; by the various white patches on the anterior wing being smaller and less numerous; and by the much larger spots of fuscous-brown which occupy the central portions of the tear, or, wedge-shaped markings on the posterior wing.

From the O. Pronomus, it differs by not exhibiting in the males, the central veins of metallic green, additional to the two curved bands on the anterior wing; by having four or five black spots, instead of three, on the posterior wing: on the underneath of the superior wing, by the much lesser green discal spot, and by the contiguous macular band being separated into two : by the females in the several markings of impure white being very much smaller and by the tear-shaped spots being of different form, with their centres more occupied by dark-brown.

From the O. Euphorion, (of which the female is only known) it differs by the faint white band in the discoidal cell and by the streak between the fourth and fifth sub-costal nervules of the primary wings not containing within it the small black spot, so minutely detailed and figured in the *Brit. Mus. Cat. p.* 4, *pl.* 2, *fig.* 3, as characteristic of the Euphorion; by the whole surface being much more obscure; and by carrying on the thorax the green longitudinal stripe—not seen in any of the specimens of the Euphorion, collected by the late Allan Cunningham,—nor in those now in the cabinet of Mr. W. MacLeay.

From the O. Richmondia, it differs by being much larger; by the males showing on the upper surface of the anterior wing more distinctly the inner marginal green border, and on the under surface a much smaller green discoidal marking : by the females possessing the tear-shaped markings more occupied by the dark-brown spots within them ; by the more general sombre appearance ; and by the green stripe on the thorax.

From the O. Poseidon (of which the male is only known) it differs by not exhibiting " the rich green colour which extends along both sides of the median nervure and partly, or, entirely along the course of the nervules of the primary wings towards the outer margin " (O. Poseidon, Brit. Mus. Cat.) ; and by having on the under-side the transverse macular band separated into two parts.

From the O. Archideus, (of which the female is only known) it differs by " the white of the secondary wings " not " reaching to the disco-cellular nervules and a portion of the median nervule," nor " occupying a small space within the discoidal cell."

And from the O. Victoriæ, (of which the female is only known) it differs by the far greater obscurity of colouring of the entire surface ; the whitish markings being in no ways proportionate to those of the Victoriæ ; indeed rendering a further comparison between them needless.

Description of new species of Articerus,

By REV. R. L. KING, B.A.

[Read 1st October, 1868.]

THE genus *Articerus* was first established by Dalman* upon a species, named by him *A. armatus*, which had been discovered in gum-copal. He was not able, however, to give any very detailed description ; nor was any thing more known of the genus until the Rev. Mr. Hope† described and figured a species sent to England from South Australia under the name *A. Fortnumi.* The next additions made to the genus were those contained in Westwood's Monograph of Australian and other *Pselaphidæ* in the Transactions of the Entomological Society of London (Vol. III., N.S., p. 271). He added *A. curvicornis, angusticollis, dila-ticornis,* and *setipes,* all from Victoria, and *A. braziliensis* from South America. Pascoe has since described a species from Western Australia, under the name *A. Bostockii,* and has distinguished the species found so abundantly in South Australia, near Gawler, by my friend Mrs. J. Kreusler, under the name *A. Odewahnii.* It is evident, however, that both these last species are remarkably close to, if not identical with the original species described by Hope, as *A. Fortnumi.* *A. Duboulayi* has been added by Waterhouse from Western Australia—a species from Syria, (*A. Syriacus*) has also been described ; and another from North America, (*A. Fuchsii*) has been added, *vide* Proceedings of Soc. Phil., 1866.

The species *Braziliensis* and *Fuchsii* appear to have been removed from the genus *Anticerus,* by Brendel, and placed under the new genus *Fustiger*; (*fustis gero*). Not having had an opportunity of consulting the diagnosis of the genus, I can only imagine that the peculiar elongate antennæ of the former species have been regarded as of sufficient importance to justify the erection of a

---

* Dalman, Om., Ins. innes i Copal, p. 23.

† Ann. Nat. Hist. XI., p. 319 ; and Trans. Ent. Soc., London, IV., p. 106, pl. viii.

new genus. In this particular, however, the species placed under the new name are certainly united with *A. Fortnumi, Bostockii*, and *Odewahnii*, and in a less decided way with *A. Curvicornis*. I prefer, therefore, for the present, to place the first of the species which I am about to describe under the old genus; although from its singular resemblance to *A. Braziliensis*, it may eventually find itself under the new genus *Fustiger*.

These insects are rare in New South Wales. But their small size, and particularly the ferocity of the ants, under whose protection they live, may in some measure account for the infrequency of their capture. I have been able, however, to add four species to our Colonial Fauna. *A. angusticollis* occurs in ants' nests, at Paramatta, and in the Liverpool Plains. *A. setipes* was captured by my son, Mr. R. King, at Goono Goono, in the Liverpool Plains district—hardly differing from specimens in my cabinet from Gawler, South Australia. *A. curvicornis* is frequently captured at Liverpool, in the nest of the small black ant—and differs only in its somewhat smaller size from the description given by Westwood, (Loc. cit.) of Melbourne specimens. The fourth species, *A. regius* (*mihi*) is as far as is yet known peculiar to Liverpool. Mr. Masters has also captured a species in debris, after a flood, at Rope's Creek, near Penrith, which I have described as *A. breviceps*.

ARTICERUS REGIUS.

Obscure castaneus, elytrorum disco pallidiori, punctatissimus, minute pubescens; capite oblongo, antennis capite longioribus linearibus cylindricis ad basin constrictis; thorace subgloboso, lateribus rotundatis; elytris sutura nigricante et linea suturali notatis; abdomine nitido parcissime setoso; pedibus robustis, tibiis maris prioribus et intermediis ad medium dentatis.

Long. mas. .14 poll.

  fem. .10 „

Ants' nests in wood; Liverpool, New South Wales, from June to September.

A specimen of the male has been deposited in the Australian Museum, and another in the collection of W. MacLeay, Esq. The correspondence between this species and *A. Braziliensis*, as described and figured by Westwood, (Loc. cit.) is certainly very close. Yet the specific differences are quite sufficient to leave no doubt on my mind that our insect is quite distinct from the American.   *A. regius* has neither the foveoles on the thorax nor the discoidal striæ on the elytra, which mark the *Braziliensis*. Westwood also describes the legs of the latter as *graciles*, a term which might apply to those of the female of *regius*, but by no means to those of the male.   In our species, the fore tibiæ of the male are deeply notched and toothed.   The intermediate legs have the tibiæ toothed at the middle, and the femur is armed with a strong spine.   The female, which is much smaller, has all the legs unarmed.

The head is slightly enlarged between the antennæ.   The antennæ are nearly straight, cylindrical, and very slightly enlarged towards either extremity.   They are somewhat longer than the head.

Westwood says of *A. Braziliensis* that it is very distinct from all the Australasian species in its sub-cylindrical antennæ, and in the form and sculpture of the head and thorax.   This discovery of our present species greatly qualifies this assertion, and adds another to an already considerable list of forms existing in the fauna of Australia closely allied to those of South America.

ARTICERUS BREVICEPS.

Brunneus setosus ; capite brevi postice rotundato, antennis capite longioribus ad apicem clavatis truncatis ; thorace ad medium valde depresso, ante medium latiori, postice subro- tundato ; elytris stria suturali notatis.

Long. .10.

Rope's Creek ; under debris after a flood.   *Mr. Masters.*

The head is very short, increasing in breadth to the eyes ; the breadth behind the eyes being nearly equal to the whole length. The antennæ are longer than the head, thin at the base, but gradually increasing in thickness for about one-half the length, and then swelling into a truncate knob; and bearing no slight resemblance to an aboriginal's "waddy."

The species is very distinct from all the other members of the genus.

The whole genus, as far as is now known, consists of the following species :—

1. *Articerus armatus.* Dalman, the type of the genus.
2. *A. (Fustiger) Braziliensis.* I. O. Westwood, S. America.
3. *A. (Fustiger) Fuchsii.* Brendel, Tennesee.
4. *A. (? Fustiger) regius* R. L. K., New South Wales.
5. *A. Fortnumi.* Hope, South Australia.
6. *A. Bostockii.* Pascoe, Western Australia.
7. *A. Odewahnii.* Pascoe, South Australia.
8. *A. curvicornis.* I. O. Westwood, N. S. W. and Victoria.
9. *A. breviceps.* R. L. K., New South Wales.
10. *A. angusticollis.* I. O. Westwood, N. S. W. and Victoria.
11. *A. setipes.* I. O. Westwood, N. S. W., South Australia, and Victoria.
12. *A. Duboulayii.* Waterhouse, Western Australia.
13. *A. Spriacus.* Saulcy, Syria.

On the Scaritidæ of New Holland, by

WILLIAM MACLEAY, ESQ., F.L.S.

[Read 6th September, 1869.]

IT is now more than four years since I wrote my third paper on the *Scaritidæ* of New Holland, and during that period large additions have, as I anticipated, been made to the list of species received from various parts of Australia. I am desirous now of adding to that list a few species which have come under my notice since the date of my last publication, but before doing so, I shall take the opportunity thus afforded me of making some observations on the genera and species of the Family described by Count de Castelnau, in the 1st volume of the Transactions of the Royal Society of Victoria, and on the excellent little treatise on the genus *Carenum*, by M. Le Baron de Chandoir, published a few months ago in the Transactions of the Entomological Society of Belgium.

The first-named of these Entomologists has, in the above cited work, described thirty-four new species of the Family, has merged my genus *Euryscaphus* in the genus *Scaraphites*, has reconstituted the genus *Eutoma* of Newman, and has formed a new genus under the name of *Neocarenum*.

The reason assigned by Count Castelnau for the rejection of the genus *Euryscaphus*, is that the character on which I founded its divison from *Scaraphites*, viz., some difference in the form of the elytra, is neither constant nor of generic importance.

A reference, however, to my description of *Euryscaphus* (Trans. Ent. Soc., N.S.W., vol. 1., page 187) will show that the Count is mistaken in supposing that I formed the genus upon any characters as distinct from *Scaraphites*, though I certainly point out the very different shape of the abdomen in the two genera. It will be found that on the contrary I point out the almost perfect identity in many respects of *Euryscaphus* with *Carenum*. In fact, the *Euryscaphi* are gigantic *Carenums*, and are

as far removed from *Scaraphites*, as are any two genera of the
Family. They differ in the head, palpi, thorax, elytra, and legs,
while in all these *Euryscaphus* nearly agrees with *Carenum*.

That there may be insects as stated by Count Castelnau,
which form an insensible passage between *Euryscaphus* and
Scaraphites, I will not deny, but I have not seen any, and the
instance cited by the Count, viz., *Scaraphites Heros*, certainly
does not from his description bear out the assertion.

But even so, I cannot admit that the discovery of a species
which appears to form a link between any two genera, is any
reason for the rejection of either of these genera.

Of the eight species of *Scaraphites*, described by Count
Castelnau in the paper above referred to, probably four species,
viz.: *Howittii, affinis, carbonarius*, and *Hopei*, belong to the genus
Euryscaphus, while the species named *Heros, humeralis, gigas*,
and *Martinii*, seem to be *Scaraphites*.

Count Castelnau appends a note to these descriptions in which
he gives it as his opinion that *Scaraphites rotundipennis* Dejean,
M'Leayi Westw., and *intermedius* mihi, are all the same species.
And here again I must complain of the Count's assertion, that I
rely entirely on the number of the marginal punctures on the
elytra for differential characters between these three species. In
page 190 of the first volume of our Transactions, I give a detailed
description of *Scaraphites intermedius*, and append to it a remark
to the effect, that a ready mode of recognizing the three species,
without the trouble of a close examination, is to count these lateral
punctures of the elytra, which though not constant in number,
seemed to be generally most numerous in *rotundipennis*, and least
so in *intermedius*.

I have no doubt myself that *intermedius* is a distinct species ;
the other two, though apparently distinct, may be merely local
varieties.

Twelve new species have been added to the genus *Carenum*,
in the Count's paper ; one of these he gives as a mere re-descrip-
tion of *C. atronitens* mihi, but it is really a new species which I
have named in my cabinet, *C. Gawlerense*, and I would suggest
that it should bear the name henceforth of *C. Gawlerense* of Castelnau, as that gentleman was the first to describe it. This species

and *C. devastator* belong to the group in my list of Scaritidæ, page 196, loc. cit., which begins with *C. quadripunctatum*. *C. Brisbanense, ebeninum,* and *Westwoodii* belong to the *C. Bonellii* group ; *C. carbonarium* and *Schomburgkii* to the *C. marginatum* group ; *C. splendens* and *Odewahnii* to the *C. coruscum* group ; and *C. multiimpressum* to the *C. Spencii* group. The other two species, *C. superbum* and *amabile,* both very remarkable insects from the Lachlan, are referred by the Baron de Chandoir, to a new genus which he has named *Conopterum,* from the peculiar shape of the elytra.

The genus *Neocarenum* has been established by Count Castelnau, on two species, *singulare* and *Kreusleri,* one of which is certainly my *Carenum elongatum,* and the other is probably only a variety. It occupies an intermediate position between the genera *Carenum* and *Eutoma.*

Of the last named genus, which will include the six species in the *C. tinctillatum* group of my list of Scaritidæ alluded to above, and the first three species of the *C. violaceum* group, Count Castelnau has re-described carefully Newman's original species, (*tinctillatum*) and described six new species, viz., *episcopale, Newmani, filiforme, purpuratum, læve,* and *Loddonense.* He has also added six species to the genus *Scarites,* viz.: *substriatus, plicatulus, Mitchellii, Bostockii, ruficornis,* and *bipunctatus.*

M. le Baron de Chandoir's "note on the genus *Carenum,*" above referred to, is deserving of special attention, inasmuch as it is the production of one who is a very high authority upon the Carabidæ generally, and who, has evidently paid great attention to the study of this particular sub-family.

He suggests in the first instance, a sub-division of the genus, founded upon more constant and reliable characters than those adopted by me in my list of the Australian Scaritidæ.

It is evident, however, that the Baron, though having a most thorough Entomological knowledge of his subject, has in his possession but few of the many species of the genus, and that therefore he is not in the best position to judge of the simplest way to render the species recognizable, which is in fact the real intent and meaning of these numerous sub-divisions.

The Baron's sub-divisions are, however, perfectly unexceptionable, and I incline to think that his Section 1, " Elytra juxta marginem costigera " including *Carenum tuberculatum* mihi and *C. carinatum* mihi (which I have no doubt is the second species alluded to by the Baron, under the name of *M'Leayi*) ought to constitute a distinct genus.

The new species of *Carenum* described in the Baron's paper, are, *C. foveigerum* of the *C. Spencii* group ; *C. transversicolle*, the position of which is somewhat doubtful ;* *C. Castelnaui*, of the *C. Bonelli* group, and probably identical with *C. interruptum* mihi ; and *C. convexum* of the *C. marginatum* group.

M. le Baron has also formed a genus upon three species, which differ chiefly from all the others of the Family, in having only one external tooth to the anterior tibiæ. This genus he has named *Monocentrum*, and in addition to *Carenum megacephalum* of Westwood, which he has referred to it, he has described two new species named respectively, *grandiceps* and *longiceps*. I have never met with any one of the three species, nor have I ever seen anything at all even remotely referable to the genus.

Under the name of *Conopterum insigne*, the Baron de Chandoir describes a peculiar form of insect also unknown to me, and it is to the same genus that he believes the species *Carenum superbum* and *amabile* of Castelnau mentioned above, should be referred. The most marked characteristics of the genus are, a large head with elytra broad at the base, and gradually decreasing towards the apex.

The genus *Carenidium* of the same author is formed on the *Carenum gagatinum* mihi. It is clearly a good genus, in the form of the elytra, somewhat resembling the last, but its most marked characteristics are its excavated labrum and pointed antennæ. As the Baron de Chandoir's treatise on the genus *Carenum* above referred to, in which this genus (*Carenidium*) is described, is not easily procurable in Sydney, I will, when I come

---

* Since writing the above, I have seen a specimen of this insect among some Coleoptera, sent by Mr. Diggles from Moreton Bay. The impunctate elytra would place it in the *C. politum* group, but the transverse rectangular thorax agrees with that of *C. rectangulare*, and is utterly unlike that of any other species of *Carenum*.

to the descriptions of two species to be added to it, give the Baron's characters of the genus in full.

One remarkable species, *Carenum mucronatum*, which I have described in the Proceedings of the Ent. Soc., N.S.W., of the 2nd October, 1865, is not noticed by either of the above named Entomologists, and from that I infer that they have never seen it. I only know of two specimens of it, one in the collection of the Rev. R. L. King, the other in my own.

The following species are I believe new :—

CARENUM SEXPUNCTATUM.

Nigrum nitidum purpureo-marginatum, elytris sexpunctatis subtilissime striatis, tibiis anticis extus bidentatis.

Long. 11 lin., lat. 3½ lin.

Hab., Lower Murrumbidgee.

This species is of the size and general appearance of *C. interruptum*, but is altogether more brilliant. Its chief peculiarity consists in having two punctures about a line apart, and parallel to the suture, on the apical third of the elytra. Its habitat seems to be the sand hills of the Riverine country.

CARENUM CYANIPENNE.

Nigrum nitidissimum sulcis frontalibus subparallelis, thorace subquadrato angulis posticis rotundatis subemarginatis, elytris subovatis convexis nigro-cyaneis subpurpurascentibus quadripunctatis, tibiis anticis extus bidentatis.

Long. 7 lin., lat. 2½ lin.

Hab., South Australia.

The only two species of the *C. Bonellii* group, to which this insect belongs, heretofore described from South Australia, are *C. anthracinum* mihi and *C. ebeninum* Casteln. The present species differs from both in being of less size and more brilliancy. The head is broad, with the facial grooves short and nearly parallel. The thorax is rather broader than long, with the anterior angles somewhat prominent, and the posterior cut away and very slightly emarginated. The elytra are oval, very smooth,

and of a beautiful dark violet tinge, with two well-marked punctiform impressions on each, one near the shoulder, the other
towards the apex.

CARENUM CHAUDOIRI.

Nigrum nitidum sulcis frontalibus divergentibus, thorace subquadrato postice subviolaceo, elytris subangustis subtilissime
striato-punctatis nigro-violaceis quadripunctatis, tibiis anticis
extus bidentatis.

Long. 9 lin., lat. 2½ lin.

Hab., Australia.

I have no record of the particular habitat of this insect, nor of
how I became possessed of the single specimen of it in my
cabinet. It is in many respects like *C. Bonellii*, but differs from
it in having the shallow transverse depression on the forehead
less marked, in the colour which has no shade of green, and in
the narrow form and more distinct sculpture of the elytra.

CARENUM OPACUM.

Nigrum opacum sulcis frontalibus divergentibus, thorace subquadrato, elytris viridi-aureis subtilissime et confertissime
punctulatis quadripunctatis, tibiis anticis extus bidentatis.

Long. 10 lin., lat. 3 lin.

Hab., Clarence River.

This species is also in many respects like *C. Bonellii*. The
facial grooves, however, are longer and deeper, and the transverse
depression behind more deep and circular. The thorax is sometimes of the same dull golden green as the elytra. These last,
under a lens, show a surface completely and closely covered with
very minute punctures, giving a shagreen appearance, and no
doubt causing the dull appearance so characteristic of the species.

CARENUM TRISTE.

Nigrum subopacum sulcis frontalibus divergentibus, thorace
subquadrato angulis posticis rotundatis, elytris ovatis quadripunctatis, tibiis anticis extus bidentatis.

Long. 8 lin., lat. 2½ lin.

Hab., Wide Bay.

This species is of a less elongate form than *C. Bonellii*, and much narrower than *C. interruptum*. It is entirely of a dullish black without any apparent marginal colouring.

CARENUM KINGII.

Nigrum nitidum violaceo-marginatum sulcis frontalibus parallelis, thorace subtranverso postice rotundato subemarginato,
elytris opacis bipunctatis, tibiis anticis extus bidentatis.

Long. 9 lin., lat. 3 lin.

Hab., Liverpool Plains.

The Rev. R. L. King has kindly lent me this insect for description. It was taken by him at Goondo Goonoo, Liverpool Plains, and is I believe unique in his collection. The dull appearance of the elytra is caused by close and minute punctura-tion, as in the case of *C. opacum* and others of the genus.

CARENUM PROPINQUUM.

Nigrum nitidum violaceo-marginatum sulcis frontalibus sub-
parallelis, thorace subtransverso antice transversim impresso
postice rotundato subemarginato, elytris subtilissime striato-
punctatis postice fortiter bipunctatis, tibiis anticis extus
bidentatis.

Long. 6 lin., lat. 2 lin.

Hab., Liverpool Plains.

The insect differs from the last, which it closely resembles, in its small size, in the transverse impression on the anterior portion of the thorax, and in the sculpture of the elytra, which are not dull, are minutely striato-punctate, and have the two punctures near the apex large and deep.

The Rev. Mr. King captured this insect in the same locality as the last described one.

.   CARENUM NITESCENS.

Nigrum nitidissimum planum capite bisulcato sulcis latis rugosis
divergentibus, thorace subquadrato angulis posticis rotundato linea dorsali fortiter impresso postice utrinque foveolato,

elytris subovatis purpureo-marginatis postice fortiter bipunctatis humeris prominentibus, tibiis anticis extus bidentatis.

Long. 6 lin., lat. 2 lin.

Hab., Salt Lake, Hummock Range, South Australia.

The flat appearance of this insect is its most striking character.
It is of a very brilliant black, with the sides of the elytra of a
violet tinge. The frontal grooves diverge a little behind, and the
face on each side of them is rugose. The thorax is as long as
broad, gradually narrowing from behind the middle to the
posterior angles which are slightly emarginated, the dorsal line is
strongly marked, and on each side of it towards the posterior
angles, there is a roundish fovea. The elytra are as broad as the
thorax, with the humeral angles advanced, and the puncture near
the apex well marked.

I am indebted to Mr. Odewahn for this as well as many other
of the South Australian species of Scaritidæ.

CARENUM INEDITUM.

Nigrum nitidum viridi-marginatum sulcis frontalibus sub-
parallelis, thorace transverso postice rotundato, elytris
impunctatis, tibiis anticis extus bidentatis.

Long. 8 lin., lat. 2¾ lin.

Hab., South Australia.

The absence of the usual punctiform impression on the elytra,
will place this insect in the same group with *C. lævipenne* mihi,
and in general appearance it is not unlike that species.

CARENUM RUFIPES.

Violaceum nitidissimum, capite nigro magno sulcis frontalibus
divergentibus, thorace transverso late marginato, abdomine
ovato, elytris seriatim punctulatis postice bipunctatis, pedibus subrufis, tibiis anticis extus tridentatis dente tertio
minuto.

Long. 8 lin., lat. 3 lin.

Hab., Stirling Range, Western Australia.

This very beautiful and distinct species is represented in the
Sydney Museum, by one rather immature specimen taken by

Mr. Masters, at the place above mentioned. *C. campestre* mihi is the species of the group which it most resembles in general form.

CARENUM SUBCYANEUM.

Nigrum nitidum viridi-marginatum sulcis frontalibus subparallelis, thorace transverso marginato postice lobato truncato utrinque emarginato, elytris nigro-cyaneis subpurpurascentibus postice bipunctatis tibiis anticis extus tridentatis.

Long. 8 lin., lat. 3 lin.

Hab., South Australia.

The above measurement is taken from the largest sized, but most perfect of five specimens in my possession. Probably 7 lines is nearer the average length. It is very different from the other species of the group in which its tridentate tibiæ will place it.

CARENUM DISPAR.

Nigrum nitidum viridi-marginatum sulcis frontalibus subparallelis, thorace transverso marginato postice rotundato vix lobato, elytris subpurpurascentibus antice latis subtruncatis postice bipunctatis, tibiis anticis extus tridentatis.

Long. 9 lin., lat. 3½ lin.

Hab., South Australia.

This also differs from the group in which my subdivision of the genus would place it, in its parallel frontal grooves, and non-oval elytra, its transverse broadly margined thorax is however quite in character with the group.

CARENUM ORDINATUM.

Nigrum nitidum late atro-viridi-marginatum, sulcis frontalibus divergentibus, thorace transverso postice rotundato sublobato, elytris ovatis seriatim punctulatis postice bipunctatis, tibiis anticis extus tridentatis.

Long. 11 lin., lat. 3½ lin.

Hab., South Australia.

The affinity of this species is to *C. Odewahnii* Castelnau; it differs from it chiefly in its darker and duller appearance, and in having seven rows of distinct but shallow punctures on each elytron.

EUTOMA MASTERSI.

Nigrum nitidum, elytris nitidissimis violaceo-marginatis leviter striato-punctatis postice bipunctatis.

Long. 6½ lin., lat. 1¾ lin.

Hab., Dabee, near Mudgee.

The punctured striæ on the elytra of this species are quite apparent under a common lens, and in this respect it may be readily distinguished from any species hitherto described.

EUTOMA DIGGLESI.

Nigrum nitidum, thorace oblongo antice truncato postice rotundato medio canaliculato transversim substriato viridi-marginato, elytris viridibus violaceo-marginatis quadri-punctatis.

Long. 6½ lin., lat. 1½ lin.

Hab., Moreton Bay. ?

I am not certain of the habitat of this species. The only specimen I have seen was sent from Brisbane by Mr. Diggles, with a number of Coleoptera, many of which were not Queensland insects.

The four punctures on the elytra of this species are remarkable, inasmuch as it is the first of the genus in which the number has exceeded two. The punctures are very large, and are placed much in the same way as in *Carenum Bonellii*.

Two very remarkable insects may here be noticed; they are probably of different and undescribed genera, but their very imperfect state, both being without antennæ or palpi, renders it impossible to give their proper characters. Their elongated form, however, and narrowly shouldered elytra seem to correspond so closely with the species " *elongatum*," upon which the

Count de Castelnau has formed the genus *Neocarenum*, that for the present I will refer them to that genus.

NEOCARENUM MASTERSI.

Nigrum subnitidum sulcis frontalibus divergentibus postice profunde antice leviter impressis, thorace elongato postice rotundato, elytris elongatis seriatim profunde punctatis ; tibiis anticis extus bidentatis.

Long. 17 lin., lat. 4 lin.

Hab., Mount Barker, Western Australia.

The enormous size and remarkably elongated form of this insect separates it at once from all others of the *Scaritidæ* known to me. The marking also of the elytra is unusual. There are nine rows of punctures on each elytron, those towards the sides are almost obsolete, but the five rows on each side of the suture are composed of close, large, deep punctures. The fore legs are very strong.

NEOCARENUM RUGOSULUM.

Nigrum subopacum elongatum capite leviter bisulcato, thorace subnitido postice subrotundato linea dorsali leviter impressa, elytris angustis ad suturam depressis leviter striatis, tibiis anticis extus bidentatis.

Long. 8 lin., lat. 2 lin.

Hab., Salt Lake, Hummock Range, South Australia.

I received a single and imperfect specimen of this curious insect, from Mr. Odewahn, a few days ago. The whole upper surface, the thorax excepted, has a slightly rugose appearance, the striæ on the elytra, though slight, are distinct, and of a wavy character. The sutural portion of the elytra is deeply indented. There is a well defined punctiform impression on the apical portion of the left elytron, but I cannot find any trace of a similar impression on the other.

Genus CARENIDIUM.   Chandoir.

Frons profunde bisulcata ; *clypeus* ad labri latera utrinque longius dentatus convexus valde declivis, inter dentes pro-

funde emarginatus, labrum amplectens.　*Caput* maximum ;
mandibulæ crassæ validæque.

Palpi maxillares (modice labiales latissime securiformes.)

Labrum parvum, antice profunde emarginatum, margine antico
declivi excavato.

Antennæ tenues, thoracis basi breviores, apicem versus attenu-
atæ, articulis septem ultimis elongato-quadratis, angustis,
valde compressis et utrinque omnino glabris, margine tantum utroque pubescente, ultimo præcedente plus dimidio
longiore, apicem versus sensim angustato, subacuminato,
summo apice piligero.

Prosternum inter coxas triangulare subexcavatum.

Episterna metasterni latitudine longiora.

Abdominis segmenta postice medio haud punctigera.

Tibiæ anticæ extus bidigitatæ intermediæ extus apice spina
longiuscula tenui armata.

Elytrorum margo tenuis (ut in *Carenis*), ad humeros haud
inflatus, usque ad pedunculum productus.

Habitus elongatus, elytris elongato-ovatis, convexis.

With two exceptions, the above is an exact copy of Baron de
Chaudoir's description of this genus.　I have inserted a description of the *Palpi*, which were wanting in the Baron's specimen,
and I have omitted the word "impunctatis," as applied to the
elytra, as I have now to describe a species which has four
punctures on the elytra, as in *Carenum Bonellii*, and yet is a
most perfect example of the genus *Carenidium*.

CARENIDIUM DAMELII.

Subnitidum supra æneo-viride subtus nigrum, elytris quadripunctatis obsolete striato-punctatis.

Long. 14 lin., lat. 4½ lin.

Hab., Cape York.

A single specimen of this fine insect was taken at Cape York,
by Mr. Damel, after whom I have named it.

CARENIDIUM KREUSLERÆ.

Nigrum subnitidum viridi-marginatum, elytris obsolete striato-punctatis.

Long. 15 lin., lat. 4 lin.

Hab., South Australia.

This species differs from *C. gagatinum* in having the labrum less deeply emarginated, the thorax and elytra deeply bordered with dull green, and these last indistinctly striated and punctured.

The only specimen I have seen, was sent to me by Mr. Odewahn, labelled "from Mrs. Kreusler, found near Gawler."

SCARAPHITES MASTERSI.

Niger nitidus, capite profunde biimpresso rugoso, thorace transverso lateribus rotundato postice truncato, elytris latis subtilissime striato-punctulatis punctis: marginalibus confertis, sub-marginalibus sex, postice suturam versus tribus impressis.

Long. 18 lin., lat. 8 lin.

Hab., Mount Barker, Western Australia.

One specimen of this fine insect was brought by Mr. Masters from Western Australia, and is now in the Sydney Museum. The species it most resembles, is the *Scaraphites Silenus*, of Westwood, but it differs from it very much in the elytra, which are more elongated in the present insect, are less distinctly striated, have the striæ finely punctured, and have three large punctures towards the apex placed parallel to the suture.

On the Byrrhides of Australia, by the
Rev. R. L. King, B.A.

[Read 22nd November, 1869.]

Although the family is not numerously represented in Australia, yet two of the genera, being peculiar to Australasia, are not without interest. The genus *Microchætes* was described by Mr. Hope, in the Transactions of the Entomological Society of London, (Vol. I., p. 153) and founded upon a species named by him *M. sphæricus*. The genus differs from the *Syncalypta* of Stephens, principally in the antennæ, of which the first joint is rather long, the second to the eighth gradually decreasing in length; the elytra are covered with tufts of stiff and generally truncate setæ analogous to those which exist in *Nosodendron*.

It is possible that future Entomologists will prefer reuniting *Microchætes* with *Syncalypta*, the comparative size of the different joints of so variable an organ as the antennæ being an unsafe character on which to rest a generic distinction. At any rate the new species which I am about to describe appears to form a passage from the one genus to the other, and at the same time to differ from them both in its tarsal developement. But that I think it probable that the genus *Microchætes* may not be retained eventually, I might have formed another genus out of my new species.

I am however under no doubt about the generic distinctness of my second new species, *Byzenia formicicola*. Its facial developement, the visibility as well as the proportions of its strange antennæ, and its curious elytra, all combine with its habits to point out a wide difference between it and the other members of the family. The remarkable forms of many of the coleoptera, which inhabit the nests of ants, have often attracted the attention of Entomologists ; our present species is no exception.

I insert, from the original descriptions, the diagnosis of the species of *Microchætes* which are already known.

Genus I. Microchætes. Hope.

In this genus the eyes, mandibles, and labrum are quite concealed when the head is retracted into the thorax. The antennæ are composed of one rather large basal joint, 2—8 gradually decreasing, and the remaining joints forming a club. All the tarsi are contractile, and received into a groove in the femur: the body is covered with tufts of short truncate setæ.

Sp. 1. *M. sphæricus.* Hope.

Totum corpus supra nigrum, fusco-tomentosum, pedibus piceis.
Long. 2 lin., lat. 1½ lin.
Swan River.

The clypeus is rounded and slightly punctured. The thorax is marked with four tubercles placed almost on the middle of the back. The elytra are bristling with tubercles disposed in a triple series—the body beneath is concolorous.

Sp. 2. *M. scoparius.* Erichs.

Niger, opacus, nigrosetosus, elytris fasciculatis.
Long. prope 2 lin.
Tasmania.

This species is described (Erichs. arch. 1842, I., p. 153) as having the body black, opaque, covered above with very fine ashy setæ. Antennæ slender, piceous. Head densely rugulose punctate, the front sprinkled with short truncate black setæ. The thorax is short, the posterior angles elongate acuminate with numerous truncate setæ on the margin. Elytra substriate with numerous truncate setæ, mostly arranged in tufts. Body underneath and feet covered with short reclined setæ.

The Australian Museum collection contains specimens of the genus from N. S. Wales, Victoria, South Australia, Western Australia, and Tasmania. I am inclined to refer the Tasmanian specimens to the latter of these descriptions, and the rest to the former; yet not without some doubt. Hardly any two are alike on the back of the thorax, and therefore I cannot lay any stress upon the four dorsal tubercles mentioned by Hope. And although there is no doubt about the number of tufts or fascicules, composed of the characteristic short truncate setæ, I do not recognize in any of these specimens the tubercles disposed in a "triple series." Several of the Australian specimens are piceous beneath the body, but not all; others are as black as the Tasmanian specimens.

In the specimens from Melbourne the tufts are more numerous and coarse. But I cannot detect any difference which requires that they should be regarded as a distinct species.

Among the Museum specimens collected by Mr. Masters at King George's Sound, there is one which probably will form a new species, but it is in a bad state for description, and whether it was originally clothed with setæ disposed in tufts, or (as in *M. minor*) in lines it is now impossible to discern. It is black and about one-half the size of those which I regard as *M. sphæricus*.

Sp. 3. *M. minor*.

Niger, elytris striato-punctatis, squamis cinereis adpressis et setis erectis truncatis longitudinailter dispositis vestitis; pedibus piceis tetrameris.

Long. .07. poll.

Paramatta, under stones in grass; rare.

Sydney; *Mr. Masters*.

The very small size and the want of *tufts* of truncate setæ at once distinguish this from the former species. The truncate setæ are placed in the punctures of the elytra, and are thus arranged in regular lines.

The antennæ are 10-jointed; the first rather long; the succeeding joints gradually decrease in length to the 5th, which is the smallest of all; the remaining joints gradually increase in breadth, though not in length, to the tenth, which is as long as the three preceding. The tarsi are all tetramerous.

Genus II. MORYCHUS. Erichs.

The genus *Morychus* of Erichsen is readily distinguished by having the anterior tarsi only concealed in the groove of the tibia; the labrum, the mandibles, and part of the eyes are visible when the head is retracted into the thorax. The elytra cover the whole of the abdomen. The genus has a somewhat wide range, viz:—from Siberia on the north to southern Africa. The detection of the genus in Australia is due to my friend Mrs. Kreusler of South Australia, from whom I have received a specimen, to which I have given the name. of—

Sp. 4. *M. heteromerus.*

Nigro-piceus, striatus, minute tuberculosus ; antennis et pedibus piceis ; tarsis heteromeris.

Long. .21 ; lat. .16 poll.

Gawler, South Australia. *Mrs. Kreusler.*

The nature of the tarsi is very remarkable, and thus, in
Australia, we have in this one small family representatives of
three of Latreille's primary divisions of the coleoptera. *Microchætes
sphæricus* is pentamerous, *M. minor* is tetramerous, and our
present species is heteromerous. Well might our late member,
the learned author of the " Horæ Entomologicæ," say, that
" absolute rules of generic distinction, founded upon minute
differences of structure, are not only faults in themselves, but
calculated to blind us altogether to those beautiful groups
which the Entomologist has so often occasion to remark in
nature."[*]

Genus III. Byzenia.

Labrum et *mandibula* semper conspicua, et antennarum articulus
 primus.

Mandibulum acutum, ad medium obsolete unidentatum.

Maxillæ bilobatæ.

Palpi labiales triarticulati *maxillares* 4-articulati, articulo ultimo
 precedente longiori.

Labrum transversum.

Antennæ 9-articulatæ, articulis 1 et 9 magnis, reliquis parvis.

Elytra brevia, totum abdomen non tegentia.

Pedes robusti.

Tarsorum articuli 1 et 5 longiores.

Corpus alatum.

Sp. 5. *B. formicicola.*

Piceus elytris gibbosis rivosis.

Long. .11. ; lat. .07 poll.

Liverpool, in nests of ants.

 This very remarkable and distinct form occurs in the nests

---

[*] Horæ Entomologicæ of W. S. MacLeay, p. 491.

of a large species of ant of the genus *Formica*. The species is readily known by its black colour ornamented on the abdomen with yellowish or bronzed setæ. The ant makes its nest in the ground under wood, rails, or logs, and the beetle is seen on the ground among the ants. Three or even four have been taken from a single nest at a time—a sultry afternoon in October—but I have never met with it elsewhere.

The head is so far retracted into the cavity of the thorax that the labrum and the mandibles and a part of the antennæ are alone visible. Of these latter organs, the first joint always, and the tip of the last joint generally, are seen. The first and the last joints are of considerable size; the first is long and broad and curved at the base; in repose it is brought down nearly to the mouth entirely concealing the eyes. The next three joints are small, the fourth being the smallest of all, 5—8 increase gradually, the ninth is nearly as long as the first, but almost cylindrical, rounded at the ends; both the first and the ninth are far larger than all the rest together. The mandibles are strong, sharp at the apex, with an obsolete denticle near the middle; the lower part is fringed with setæ. The maxillæ are small and bilobed. The labrum is transverse and ogee-shaped. The thorax is *very* transverse. The elytra are short, leaving exposed the last two joints of the abdomen. They are marked by four strong ridges all rising towards one point, and nearly meeting behind the shoulder, the apparent perforation between the points being fringed with a few stiff yellowish setæ. This peculiarity of formation gives the insect the appearance of being transversely divided nearly in the centre. The scutellum is small. The tarsi are all concealed in repose in grooves in the tibiæ, but the legs which are robust are not received into cavities, as in *Microchætes*. All the legs are very widely separated.

It is not easy to trace the affinities of this remarkable form. There is no question that it belongs to the family of the *Byrrhidæ*, notwithstanding the shortness of the elytra leaving the abdomen partly uncovered, and the great distance of all the feet from each other—particulars in which it is distinguished, I think, from all the other members of the group. The nine jointed antennæ are also peculiar to itself alone of all the *Byrrhidæ*.

Description of Hiketes, a new genus of Formicicolous Coleoptera,
By the REV. R. L. KING, B.A.

[Read 22nd November, 1869.

IN my frequent inspection of ant's nests in the neighbourhood of Liverpool, I have frequently met with an undescribed form in some particulars strikingly anomalous. As a contribution to an account of our numerous Coleoptera inhabiting the nests of ants, I wish to describe it here. I have named the genus *Hiketes*, and offer the following as the diagnosis.

HIKETES. *nov. gen.*

(ʹικετης. *a supplicant.*)

Mandibulum truncatum parvum.

Mentum transversum, lobis lateralibus obtusis.

Submentum elongatum, postice liberum.

Palpi labiales 2-articulati.

Maxillares 3-articulati.

Maxillæ ?

Oculi parvi.

Antennæ ante oculos positæ 9-articulatæ, 1mo reliquis longiori, 2—7 subæqualibus, 8 et 9 clavem formantibus.

Thorax subrotundus, lateribus serratis.

Elytra abdomen tegentia.

Abdomen 5-articulatum, segmento primo majore.

Pedes priores et intermedii contigui—posteriores distantes ; femoribus magnis.

Tarsi pentameri, articulis rotundis, 4to minori.

The genus, which I have thus described, presents several very remarkable peculiarities. The parts of the mouth are very small, almost rudimentary ; and are placed at the extremity of the head, protected by the emarginate clypeus. The antennæ are short and stout, placed near the mouth, and far in advance of the eyes.

The eyes are protected by being sunk in a deep groove formed by prominent ridges on the sides of the head. Behind the mentum is an oblong plate, *the posterior portion of which is quite free,* the end emarginate. I have called it in the above description the *submentum.*

The nearest approach to it which I have met with is in Newman's genus *Deratuphrus* (or *Sigerpes* of Germar.) I have not however seen Newman's description which has been founded on four New Holland species, (The Entomologist p. 403.) In *Hiketes* the submentum is placed longitudinally, not as in *Deratuphrus* transversely. In the latter genus also the antennæ are 11-jointed—the mandibles are very robust—the thorax much elongated, the body subcylindrical and the tarsi tetramerous—all points of contrast with our present species. There can however be little doubt that the two genera are very closely allied ; and that notwithstanding its pentamerous character our genus must—for the present—take its place among the *Colydiens* of La Cordaire, and come next to *Deratuphrus* among the sub-tribe *Bothrideridæs.*

Sp. 1. H. costatus.

Castaneus punctatus ; thorace 5-costato ; elytro 3-costato.

Long. ⅓ poll.

Liverpool, in ant's nests.

The whole surface is covered with deep and wide punctures. The head is flat ; the two ridges forming the groove in which the eye is placed meet in front of that organ, the under ridge being serrated. The thorax is slightly convex, marked by 5 prominent longitudinal costæ, and has the sides rounded and serrated. Each elytron has three longitudinal costæ. The scutellum is large and punctured. The submentum extends from the mentum to the eyes. It is nearly twice as long as broad, deeply punctate. The first segment of the abdomen is wider than the rest, and separates, by a triangular plate, the coxæ of the posterior legs. The femurs of all the legs are very wide—nearly as broad as long. The tibiæ of the male are produced into a sharp point beyond the insertion of the tarsi ; in the female the termination of the tibiæ is obtuse.

This interesting species is found in the nest of a small red ant *(formica)* living in wood and under bark of dead trees on the ground.

I have much pleasure in adding a second species detected by my friend Mr. Masters, at King George's Sound.

Sp. 2.   *H. thoracicus.*

Rufo-castaneus ; thorace utrinque dilatato, costis quatuor rotato.

Long. .14.

King George's Sound—in ant's nests.   *Mr. Masters.*

The head is more produced than in the preceding species, and coarser at the sides.   The eyes are placed in a deep notch, and are visible both above and below.   The thorax is marked with four costæ, and is remarkably developed at the sides so as to resemble the wings of a Ray.   This very distinct species was found in the nest of a large black ant, under a stone at King George's Sound, in December, 1869.

Notes on a collection of Insects from Gayndah, by
WILLIAM MACLEAY, Esq., F.L.S.

[Read 3rd April, 1871.]

MR. MASTERS, the assistant Curator of the Australian Museum, has lately returned from Gayndah, a town on the Burnett River, about 150 miles inland from Wide Bay, where he had been employed for some months in endeavouring to procure for the Museum, specimens of the new description of Fish or Batrachian, lately described by Mr. Krefft under the name of *Ceratodus Forsteri.*

Mr. Masters has not only been so successful in the object of his mission as to get nineteen of these anomalous animals, but has also brought back with him a very large collection of specimens in all branches of Natural History. Among these the collection of Coleoptera stands pre-eminent, it contains over 1,100 species, and numbers nearly 16,000 specimens.

I propose to give as far as I am able in this paper, a complete list of this very magnificent collection, describing the new genera and species, and making occasional observations on the habitats &c. of the others.

I have always hitherto in describing new genera and species, adopted the system most usual with English Entomologists of giving these descriptions in Latin. On this occasion I intend to depart from that rule, as I believe that many of those who take an interest in Australian Entomology, will infinitely prefer the descriptions given in plain and intelligible English.

CICINDELIDÆ.

1.—TETRACHA CRUCIGERA, MacL., W. *Trans. Ent. Soc. N. S. Wales,* 1863, *Vol.* 1, *page* 10.

I described this species from specimens from Rockhampton and Port Denison. It is not as suggested by Count Castelnau *(Not. Aust. Coleopt., page 3),* identical with *T. Australasiæ,* Hope, which is from Port Essington.

F

2.—CICINDELA CIRCUMCINCTA, Casteln. *Not. Aust. Col.*
1867, *page* 4.

3.—DISTYPSIDERA UNDULATA, Westw. *Mag. Zool., Vol.*
1, *page* 252.

4.—DISTYPSIDERA MASTERSII. n. sp.

Length 5 lines, width 1½ lines.

This species is of a bronzy olive hue above, with the middle
of the labrum, the basal joints of the palpi, the under side of the
thighs, the tibiæ, the tarsi, an arcuated fascia extending from
the shoulders to near the suture at one third from the base, a
zigzag fascia about the middle enlarged at the sides and not
reaching the suture, and the apex of the elytra, of a pale yellow.
The antennæ, excepting the basal joint, and the apex of the
tibiæ, and of each joint of the tarsi, are black. Between the eyes the
head is marked with a number of fine longitudinal striolæ, on
the back of the head the striolæ are transverse. The thorax
is a little longer than the breadth, and is marked with
transverse striolæ ; the median line is distinct, and the transverse
depressions near the apex and base are very deep. Between
these depressions the sides of the thorax are slightly rounded.
The scutellum is short, broad, depressed in the middle and
pointed at the apex. The elytra are a little broader than the
thorax, square at the shoulders, parallel-sided, and round at the
apex. The sculpture consists of, fine wavy transverse striolæ on
the middle, and punctures on the sides. The legs are thinly
clothed with short white setæ. The under surface is of an uniform brilliant bluish black.

All the species of this genus very much resemble one another,
and a mere description without comparison would scarcely suffice
for the certain recognition of any one of them. In this case the
strongest resemblance is to *D. undulata* Westw., it differs in being
very much smaller, in having the indentations on the sides of
the labrum much deeper, in the flatter and less laterally rounded
thorax, in the smoother and less deep sculpture of both thorax
and elytra, in the shorter scutellum, in the absence of a smooth

protuberance on the shoulder and in the legs being clothed with shorter and more thinly distributed white hairs. To *D. volitans*, MacL., W., there is a near approach in size and sculpture, but in this last species the colour is darker, the yellow bands are differently disposed, the labrum is distinctly nine-toothed whereas seven only are visible in *D. Mastersii*, the scutellum is much longer and the legs are much more densely clothed with hairs. In almost all the points of difference mentioned above, it differs still more from *D. flavicans*, Chaudoir. *D. Gruti* Pasc, from Lizard Island, the only other Australian species of the genus described, I have never seen.

CARABIDÆ.

5.—PAMBORUS VIRIDIS, Gory. *Mon. t.* 161. *f.* 1.

6.—PAMBORUS GUERINII, Gory. *Mag. Zool.*, 1830, *t.* 26,—*Mon. t.* 167. *f.* 2,—*Boisd. Voy. Astrol.* 2, *page* 27.

The insect before me is evidently the small and black variety of *P. Guerinii*, alluded to by Count Castelnau in his notes on Australian Coleoptera, as coming from the Pine Mountains, Queensland.

I believe it will be found to constitute a distinct species.

7.—PAMBORUS BRISBANENSIS, Casteln. *Not. Aust. Col.* 1867, *page* 10.

8.—CASNONIA OBSCURA, Casteln. *Not. Aust. Col.* 1867, *page* 14.

9.—EUDALIA LATIPENNIS, MacL., W. *Trans. Ent. Soc. N. S. Wales, Vol.* 1, *page* 108. *Casteln, Not. Aust. Col., page* 16.

This insect was originally described by me as *Odacantha latipennis*. Count Castelnau has placed it, and properly, in a new genus, which he has named as above, but unfortunately without giving the generic characters.

10.—DRYPTA AUSTRALIS, Dej.　　*Spec. Coleopt., Vol.*
1, *page* 185.

11.—DRYPTA MASTERSII.　n. sp.

Length 4 lines.

This beautiful Drypta is very distinct from *D. Australis,* the only Australian species hitherto known.　The palpi are more pointed, the thorax has the median line much less distinct, the elytra have the lateral margins, apex, and sutural fascia, of a more brilliant blue, and are altogether more convex and strongly punctured.　The sutural facia, which is very broad and at the base extends to the shoulders, terminates at one third from the apex, where it is nearly met by a narrow extension upwards on each elytron of the blue apical margin.　There is a dense ashen pubescence over the entire surface of the body.

12.—POLYSTICHUS AUSTRALIS.　n. sp.

Length 4 lines.

Of a rather dull brown covered with a dense short yellow pubescence, with the antennæ, palpi, and legs reddish yellow. Head smooth and thinly punctured.　Thorax a little longer than the breadth, and slightly narrowed behind, with the posterior angles rather sharp and with a broad recurved lateral margin, on which are two setigerous punctures, one above the middle and at the broadest part of the thorax, the other at the posterior angle. Scutellum long, triangular, and densely punctured on the basal portion.　Elytra closely covered with narrow striæ, with the interstices elevated, every third interstice being distinctly larger ; of these striæ there are about thirty on each elytron, the lateral one only is broad and marked with large and rather distant punctures.

I believe this to be the first of this genus described as Australian.　I have two specimens of it in my own collection from Rockhampton.

13.—GIGADEMA POLITULUM. n. sp.

Length 12 lines.

Black. Head and thorax nitid and thinly punctured, the former elongated and narrowed behind the eyes, the latter, with the anterior angles much rounded and advanced, the apex truncated, and median line well marked. Elytra closely covered with small punctures, and striated, with the interstices elevated.

This species though smaller, most resembles *G. longipenne* Germ., it however differs from it in many respects, more especially in the long neck and in the rounder and more prominent anterior angles of the thorax. The apex of the thorax also is in this species quite straight, while in the other it is a little concave.

14.—GIGADEMA SULCATUM, MacL. W.

Helluo sulcatus, MacL., W. *Trans. Ent. Soc. N. S. Wales*, 1, 1864, *page* 108.

Gigadema Thomsoni, Casteln. *Not. Aust. Col.* 1867, *page* 21.

15.—HELLUOSOMA MASTERSII. n. sp.

Length 6 lines.

Entirely of a reddish brown colour. Head very much constricted behind forming a small neck. Thorax elongated, truncate in front, much narrowed behind, covered with coarse punctures and longitudinal depressions bounded by rough irregular ridges, with the median line extremely fine and in the middle of the largest of these depressions. Elytra striated and closely punctured. Legs and under side of body, abdomen excepted, of a lighter colour than the back.

Drs. Gemminger and Harold have in their " Catalogue Coleopterorum," merged the genera *Gigadema*, Thoms. and *Helluosoma*, Casteln., in the genus *Enigma*, Newm. I have, however, preserved them distinct here, as I believe them to be all good genera.

16.—ACROGENYS HIRSUTA, MacL., W.   *Trans. Ent.
Soc. N. S. Wales, Vol.* 1, *page* 109.

17.—PHEROPSOPHUS VERTICALIS, Dej.   *Spec.* 1, *page*
302.

18.—CALLEIDA PALLIDICOLLIS, MacL., W.   *Trigonothops
pallidicollis,* MacL., W.   *Trans. Ent. Soc. N. S.
Wales, Vol.* 1, *page* 110.

This may possibly be a new species, the head looks more
smooth, and the thorax narrower than in the specimens I
originally got from Port Denison.   Mr. Masters only captured
one specimen.

19.—XANTHOPHŒA CHAUDOIRI.   n. sp.

Length 4 lines.

Pale red with the elytra still paler.   The thorax is rather elongate, very little broader in front than behind, and about the
middle, where it is a little widened, there is a setigerous puncture.
The median line is deeply marked.   The elytra are striated,
with the striæ of a dark hue, and closely punctured.

I have placed this species under the genus *Xanthophœa,*
because it answers to the description of that genus, but the
thorax in *X. grandis* Chaud. is more cordiform, and that insect has
altogether a more elongate look.   I have named the species after
the founder of the genus, the Baron de Chaudoir.

20.—CYMINDIS CRASSICEPS.   n. sp.

Length 4 lines.

Of a deep black, with the exception of the antennæ, palpi,
tibiæ, and tarsi.   Head broad, large, convex, and entirely and closely
covered with longitudinal striolæ.   Thorax very little narrowed
behind, median line well marked, lateral margin furnished with
two setigerous punctures,—one about one-third of the length of
the thorax from the anterior angles, the other at the posterior
angles,—and upper surface marked with small punctures, and

fine transverse striolæ. The elytra show a slight bronzy tint, with the striæ deep and finely punctured and with the interstices elevated and smooth.

PHLŒOCARABUS. n. gen.

Head small, obtuse in front, narrowed suddenly behind the eyes into a distinct neck. Mentum with a large obtuse median tooth. Labium rather long, rounded at the apex. Labial palpi, moderately securiform; maxillary, cylindrical. Mandibles short, broad at the base, acute and slightly arcuated at the tip. Labrum almost square. Antennæ of medium length, filiform, the first joint largest, the second very small. Thorax rather broader than the length, lobed behind. Legs of medium size, fourth article of the tarsi, small and entire.

In many respects the single species of which this genus is formed resembles *Cymindis*, while in others it seems to approach *Lebia*. I prefer to associate it with the first named genus.

21.—PHLŒOCARABUS MASTERSII. n. sp.

Length 3¾ lines.

Head and thorax of a dark red, the first almost smooth, the latter with fine transverse striolæ, and well marked median line. The elytra are black, with a large spot at the base on each side of the suture, and a narrow lateral margin of a dull red. The legs and under portion of the body are of a lighter colour.

PHLŒODROMIUS. n. gen.

Head almost square, very obtuse in front, not narrower behind the eyes. Mentum without median tooth, the lateral lobes large and rounded, forming internally a right angle with the body of the mentum. Palpi short and slightly truncated. Labium oblong, truncated and covered with hairs. Labrum broader than long, widening gradually towards the apex, which is truncate. Mandibles strong, short, thick and slightly arcuated at the extremities. Antennæ short, first joint large, second small and thin, third also thin but larger than second, fourth short obconic, fifth, sixth, seventh and eight joints much thicker and somewhat square or moniliform, the remaining joints filiform. Thorax broader than the length, truncate in front and at the base, rounded at the anterior angles where it is rather narrower than at the posterior, the latter slightly obtuse. Elytra much broader than the thorax. Legs robust, fourth joint of the tarsi strongly bilobed.

22.—Phlœodromius piceus. n. sp.

Length 4½ lines.

Pitchy red, of a lighter hue on the head, sides and centre of thorax, legs and underside of body. The head and thorax are almost smooth or very remotely and finely punctured, the latter is broadly margined, more especially towards the posterior angles which are recurved. The median line is pretty well marked, and on each lateral margin there are two setigerous punctures, one at the widest part of the thorax which is almost the centre, the other very near the posterior angles. The elytra are striated, the striæ are closely and finely punctured, and the interstices are broad and rather flat.

I am somewhat puzzled as to the position of this insect, it is so unlike any of the *Lebiidæ* hitherto described, but that its effinity is to that Family, and as I think to no other is nearly certain. Mr. Masters brought one specimen only from Gayndah, but I find that I had previously received a few specimens from other portions of Queensland.

Eulebia. n. gen.

Head oval, scarcely narrower behind than in front of the eyes. Mentum with an apparently small median tooth. Palpi cylindrical and rather blunted at the apex. Mandibles strong, short, slightly arcuated at the tip. Labum square, truncated at the apex. Labium almost corneous, slightly rounded in front. Antennæ rather short, nearly filiform, the joints from the fourth to the ninth rather thicker. Thorax much broader than the length, not narrower behind, the apex truncate, the base extended in the middle into a lobe. Elytra broader than the thorax. Legs rather long, the fourth joint of the tarsi strongly bilobed.

23.—EULEBIA PLAGIATA. n. sp.

Length 3 lines.

Reddish testaceous with a large patch on the elytra extending from the base on each side of the suture, to a short distance from the apex, and the terminal segments of the abdomen of a bluish black. The head is smooth. The thorax is very finely and transversely striolate, with the median line well marked, and two setigerous punctures on each lateral margin, one a little behind the anterior angle, the other very closely above the posterior angle. The elytra are striated, with the interstices wide and slightly raised.

24.—EULEBIA PICIPENNIS. n. sp.

Length $2\frac{1}{4}$ lines.

Entirely of a pitchy red colour, rather darker on the elytra. Head and thorax as in the last species. Elytra more deeply striated, with the interstices more elevated, and with a depression on the third interstice, a little above the middle.

25.—SAROTHROCREPIS MASTERSII. n. sp.

Length 3 lines.

This species though much smaller, looks very like *Lebia posticalis* Guer. It is however of rather a paler hue, and the black fascia on the elytra is larger.

26.—SAROTHROCREPIS PALLIDA. n. sp.

Length $2\frac{1}{2}$ lines.

Smaller than the last, and of a still paler yellow. There is a large black patch about the middle of the elytra, from which there is a narrow extension behind to the lateral margins. There is also a very narrow black vitta along the suture. The labrum is not so extended as in the last species.

27.—SAROTHROCREPIS FASCIATA. n. sp.

Length 2 lines.

Like the last but much smaller, and with a broad black fascia on the hinder part of the elytra which is prolonged along the suture towards the apex.

28.—SAROTHROCREPIS MINIMA, MacL., W. *Trans. Ent. Soc. N. S. Wales*, 1864, *page* III.

This insect ought to form a new genus. I have not been able to spare a specimen for dissection, but I am convinced I was wrong in putting it with *Sarothrocrepis*.

29.—DROMIUS HUMERALIS. n. sp.

Length 1¾ lines.

Glossy, black, with the antennæ, palpi, legs, a large irregular spot on each shoulder covering nearly the whole of the base of the elytra, and a round spot on the suture near the apex, of a dark red. The head and thorax are smooth, the elytra are very lightly striated, and very minutely punctured.

There is only one specimen of this insect in the collection; I have been unable therefore to make a very minute examination of it, but I have little doubt that it belongs to this genus.

30.—HOMETHES VELUTINUS. n. sp.

Length 3½ lines.

Of a deep velvety black, with the apical half of the antennæ and the legs pale yellow. The thorax has its broadest part near the middle where there is a setigerous puncture, the median line is indistinct. The elytra are lightly striated, and have a series of depressed points along the third interstice and near the sides; these depressions present a slightly yellow appearance.

The shape of the thorax at once distinguishes this species from *H. elegans*, Newm., it is in this insect shorter and more regularly rounded, bringing the broadest part close to the middle, while in *H. elegans* the broadest part is near the apex, the sides of the thorax are not rounded from that to the base

31.—HOMETHES MARGINIPENNIS. n. sp.

Length $3\frac{1}{4}$ lines.

Black, opaque, with the antennæ, legs and lateral margin of the elytra pale yellow. The head is smooth. The thorax is shagreened, with the broadest part in the middle of its length, where it is on each side angulated and furnished with a setigerous puncture. The elytra are lightly striated, the striæ have an interrupted appearance, along the yellow margin there is a series of irregular depressions.

H. guttifer Germ. is not unlike this species, but the smooth elytra present a great contrast.

32.—PHILOPHLŒUS UNICOLOR, Chaud.   *Ann. Soc. Ent. Belg. tome* 12, *page* 220.

33.—PHILOPHLŒUS MACULATUS. n. sp.

Length 4 lines.

Head and thorax of a pitchy red, the elytra brown with a small elongated yellow spot in the centre of their basal half, and a small round spot at the apex. The sides of the thorax are angulated in the middle, where there is one setigerous puncture, there is another at the posterior angle. The elytra are very finely punctured, sulcated very shallowly and covered with fine short yellow pubescence.

34.—PHILOPHLŒUS BRUNNIPENNIS. n. sp.

Length $3\frac{1}{4}$ lines.

Red, with the elytra brown. Labrum with a well marked longitudinal stria. Thorax broad, broadly margined and punctured. Elytra broad, short, very finely punctured and lightly striated. Under side of body and legs reddish yellow, with the exception of the terminal segments of the abdomen which are black.

35.—PHILOPHLŒUS DUBIUS. n. sp.

Length 2¾ lines.

The head is black and smooth, with the antennae and palpi red, the labrum with a well defined median line towards the apex, and the penultimate joint of the maxillary palpi shorter than the terminal one. The thorax is like that of the last species, except that the setiform puncture marking the broadest part, is in this case nearer the apex than the base. The elytra are striated with a row of fine punctures in the striæ; the interstices are flat and smooth ; the sides are very narrowly margined with red, and there is on each elytron near the apex, a narrow, very zigzag fascia of the same colour extending from the suture to the seventh stria.

This perhaps ought properly to be considered an *Agonocheila*, or it may form a new genus. It resembles the last described species in having the labrum with a distinct median line. With this exception which I have not observed generally in the genus *Philophlœus*, the labrum is long and rounded as in that genus, while the palpi are entirely of the character of Baron de Chaudoir's genus, *Agonocheila*.

36.—PHILOPHLŒUS VITTATUS. n. sp.

Length 2¾ lines.

Head, thorax, and legs reddish, the elytra brown, with a yellow vitta on each extending from the base to the apex, but very much narrowed behind the middle. The labrum which is not quite so long as in the typical species of the genus, is deeply impressed in the middle, the last joint of the tarsi is longer than the preceding one, the head is smooth, the thorax sparingly punctured, the elytra very finely punctured, lightly striated as in the genus generally, with two setigerous points on the third interstice. There is a very narrow yellow margin to the elytra.

According to the definition of the Baron de Chaudoir, this ought certainly to be classed with *Agonocheila*. It resembles, however, in so many ways the genus *Philophlœus*, that, I prefer to place it with the latter.

37.—AGONOCHEILA SUTURALIS. n. sp.

Length 2 lines.

Reddish yellow with the fore part of the head, a sutural fascia on the elytra extending from the base to near the apex, and a lateral fascia on the same considerably enlarged near the apex. black. The thorax is very broad and sparingly punctured. The elytra also are broad, very closely and finely punctured, very lightly furrowed, and covered with a very fine short yellow pile.

EUCALYPTOCOLA. n. gen.

Head not narrowed behind the eyes. Mentum with a large acute medium tooth. Labium, somewhat long and rounded at the apex. Palpi, sub-cylindrical, obtuse. Mandibles, slight, long, acute, and only arcuated at the very tip. Labrum, very long, rounded, and slightly narrowed at the apex. Antennæ of medium length, thicker from the fifth joint. Thorax broader than the length, not narrower behind than in front, slightly rounded at the sides. Elytra broader than the thorax, rounded laterally and sinuate-truncate behind ; Legs strong, tarsi filiform, fourth joint small. entire.

38.—EUCALYPTOCOLA MASTERSII. n sp.

Pitchy black with the antennæ, palpi, legs, margins of thorax and elytra, and two broad and very irregular fasciæ on the latter : consisting of elongated spots on the interstices between the striæ, one above the middle, the other at the apex ; yellow. The head and thorax are nearly smooth. The elytra are sharply striated, with the interstices broad, smooth, and rather elevated.

There is an evident resemblance to the genus *Thyreopterus* in this insect, but the true affinity is no doubt to *Philophlœus*.

I have seen specimens of the genus if not the species from the Clarence River District.

39.—SCOPODES ÆNEUS. n. sp.

Length 2¾ lines.

Above brassy black. beneath bluish black, with the base of the antennæ, the palpi, and the tibiæ, yellow. The head is closely covered with longitudinal striolæ. The thorax has the same sculpture, but with the striolæ in all directions, giving it a very rugose appearance, its surface is flat, and not canaliculated, with the apex and base truncated sharply, and with the sides, angled and furnished with a long seta at its widest part which is above the middle. The elytra are opaque, with a slight silvery reflection in patches, they have several irregular depressions on their surface, and are covered with rows of small rather distant punctures, each occupied by a silvery white scale.

40.—SCOPODES LAEVIS.   n. sp.

Length 2 lines.

Glossy black, eyes very large, thorax smooth, canaliculate, widened in the middle where it forms an angle and is furnished with a long seta, and much narrowed at the base with the posterior angles rounded. The elytra are finely striated with distinct punctures in the striæ, the interstices are flat and broad.

41.—SCOPODES ANGULICOLLIS.   n. sp.

Length 2 lines.

The sculpture of the head and thorax is the same in this species as in S. æneus. But the thorax is broad, rounded at the anterior angles, and very slightly narrowed at the posterior angles, which are sharp and recurved, while on each side there are two setigerous punctures, one at the widest point a little below the anterior angles, the other at the posterior angles. The elytra are lightly striated, the striæ are rather irregular and crooked, the interstices are broad and rounded.

The thorax of this species is quite like that of a Lebia in form.

42.—SCOPODES AURATUS.   n. sp.

Length 2 lines.

Of a bronzy hue, with patches of golden yellow, especially on the elytra, and with the antennæ, palpi and legs reddish. The head and thorax are smooth but not glossy, the latter is canaliculate, and is shaped much like that of the last described species (*S. angulicollis*), with the exception of the anterior angles which are square and sharp. The elytra are profoundly striated, and there are several irregular depressions over their surface.

43.—SCOPODES SERICEUS. n. sp.

Length 1¾ lines.

Brassy black with a silken gloss on the elytra, legs and antennae brown. The thorax has a shagreen appearance, is finely canaliculated, is largely rounded at the apex and base, and has both anterior and posterior angles very sharp and prominent. Each angle is furnished with a long seta. The elytra are opaque and striated, the interstices are somewhat rounded. There are three shallow impressions along the third interstice.

44.—SILPHOMORPHA SPECIOSA, Pasc. *Journ. Ent. Vol.* 2, *page* 26.

45.—SILPHOMORPHA VICINA, Casteln. *Not. Aust. Col.* 1867, *page* 28.

I am not quite sure of the identity of this insect with the one described by Count Castelnau.

46.—SILPHOMORPHA POLITA. n. sp.

Length 5 lines.

Of a short oval form, with the upper surface of a glossy black, without puncturation or marking, and with only faint traces of striae on the elytra. The under surface is of a pitchy red.

I cannot find anything that answers to this insect among the species of this genus described by Count Castelnau, but there are so many of them almost exactly alike, that I am not at all confident as to its being undescribed.

47.—SILPHOMORPHA BIPLAGIATA, Casteln. *Not. Aust. Col.* 1867, *page* 26.

48.—SILPHOMORPHA PICTA, Castelu. *Not. Aust. Col.*
1867, *page* 25.

49.—SILPHOMORPHA BICOLOR, Castelu. *Not. Aust. Col.*
1867, *page* 25.

50.—SILPHOMORPHA HYDROPOROIDES, Westw. *Mon.*
page 401, *t.* 14. *f.* 3.

51.—SILPHOMORPHA QUADRIMACULATA, MacL., W.
Trans. Ent. Soc. N. S. Wales, Vol. 1, *page* 113.

52.—SILPHOMORPHA RUFO-MARGINATA. n. sp.

Length 3½ lines.

Of a glossy black with the margins of thorax and elytra, and
a rather broad sutural vitta on the latter, dull red. There are
two rather deep impressions near the base of the thorax, and a
few shallower irregular ones near the base of the elytra.

This species most resembles *S. suturalis* Germ., it is however
very much smaller and more convex.

53.—SILPHOMORPHA MACULIGERA, MacL., W. *Trans.*
Ent. Soc. N. S. Wales, Vol. 1, *page* 113.

54.—ADELOTOPUS MASTERSII. n. sp.

Length 5 lines.

Entirely of a rather glossy black. The elytra which are of
the same width as the thorax at the base are lightly striated, and
have a ridge on the suture near the apex.

This species much resembles *A. gyrinoides,* Hope. It is a
large broad convex insect entirely without puncturation.

55.—ADELOTOPUS BIMACULATUS, MacL., W. *Trans.*
Ent. Soc. N. S. Wales, Vol. 1, *page* 113.

56.—ADELOTOPUS SUB-OPACUS. n. sp.

Length 2¾ lines.

Dark brown with the middle of the elytra of a dull red. The

head is vertical and closely punctured. The thorax is also closely punctured, with a faint median line; it is broad and rounded at the anterior angles, where the reflexed margin is very wide, and there are two large depressions at the base nearer the posterior angle than the centre. The elytra are somewhat narrower than the thorax at its widest part, and are marked with nearly obsolete ridges and coarse punctures.

This species evidently belongs to the same group as *A. Ipsvides*, Westw.

57.—ADELOTOPUS CASTANEUS, Casteln. *Not. Aust. Col.* 1867, *page* 33.

58.—ADELOTOPUS ANALIS. n. sp.

Length 1¾ lines.

Glossy black, with the apex of the elytra red. The whole upper surface is covered with fine punctures and short erect light coloured hairs. The thorax is rounded at the posterior angles, with the margin wide, but not much reflexed.

59.—ADELOTOPUS MACULIPENNIS. n. sp.

Length 1¾ lines.

Very nitid, black, with a tinge of dark red on the thorax, and a large spot in the centre of the elytra of the same hue. The thorax is very narrowly margined and nearly cylindrical.

60.—APOTOMUS MASTERSII. n. sp.

Length 1½ lines.

In colouring, this species exactly resembles *A. Australis*, it differs from it in having the thorax more globular, the median line scarcely marked, and the posterior lobe much shorter. The elytra are striated, with the striæ full of large punctures.

61.—MORIO LONGICOLLIS. n. sp.

Length 6½ lines.

Black, subnitid, the head smooth in front and deeply im-

G

pressed on each side, with the clypeus rough and deeply marked. The thorax is longer than the breadth, is slightly narrowed at the posterior angles which are acute, and has the median line, and the posterior, and transverse impressions well marked.   The elytra are deeply striated, with the interstices convex.

This insect differs from *M. Australis* in the deeper grooves on each side of the head, in the marks on the clypeus, and in the comparatively greater length of the thorax.

62.—MORIO SETICOLLIS.   n. sp.

Length 4 lines.

This species is of a more elongate form than the last.   The clypeus seems quite smooth, and the head is less deeply grooved than in *M. longicollis*.

63. —SETALIS NIGER, Casteln.   *Not. Aust. Col.* 1867, *page* 40.

64.—VERADIA BRISBANENSIS, Casteln.   *Not. Aust. Col.* 1867, *page* 40.

PHILOSCAPHUS.   n. gen.

Body without wings.   Head large and bisulcate.   Palpi enlarged towards the apex and truncate.   Mentum with acute median tooth.   Antennæ slight, filiform, and pointed at the tip. Thorax transverse.   Elytra ovate, with a strong costiform margin, and one or two large grooves on the sides beneath.   Anterior tibiæ tridentated externally.

In addition to the species which I name below from Gayndah, this genus will comprise *Carenum tuberculatum and carinatum* named by me, and two others undescribed in the collection of the Australian Museum, one from South Australia, the other from Nicol Bay.

65.—PHILOSCAPHUS MASTERSII.   n. sp.

Length 13 lines.

The general resemblance of this species to *P. tuberculatum* is very great, it can be, however, easily distinguished by the thorax being marked with deeper transverse rugæ, and with a deeper emargination in the centre of the posterior lobe. The elytra also are much rougher and more largely tuberculate, with an interrupted tuberculate costa extending from the inner side of each humeral angle to beyond the middle of each elytron. The two other species alluded to as being undescribed differ in having an additional lateral costa, forming two deep lateral grooves, instead of one as in this species and in *tuberculatum*. *P. carinatum* is different in so many respects that comparison with it is unnecessary.

66.—Carenum salebrosum. n. sp.

Length 8½ lines.

In general appearance not unlike *C. loculosum*, Newm., but differing very much in the sculpture of the elytra. Each elytron is marked with three irregular and crooked ridges, one near the suture, another more elevated near the centre, and the third between that and the lateral margin, the most irregular of all and scarcely traceable beyond the middle. Between the suture and the first of these ridges, the space is closely filled by deep transverse impressions, and between that and the second ridge there is a series of large deep and square foveæ, all the rest of the elytra are covered with large deep foveæ of the most irregular form.

67.—Carenum occultum. n. sp.

Length 11 lines.

This species very much resembles *C. interruptum*, it is, however, of a more elongate form, and has a narrow border of a pale blue colour, very different from the brilliant purple of the other.

68.—Carenum viridi-marginatum. n. sp.

Length 9 lines.

The resemblance of this insect to *C. marginatum* is very great; the elytra are, however, proportionately shorter and more ovate, and the basal lobe or prolongation of the thorax is distinctly angular at the sides and emarginate in the middle.

69.—CARENUM POLITULUM. n. sp.

Length 7½ lines.

This species is also like *C. marginatum*, but the elytra are without trace of striæ, and are of a more elongate character. The basal lobe of the thorax is emarginate in the middle, but not angular at the sides as in the last described species. The whole insect is of a glossy black tinged with blue on the elytra, and with a green margin to both thorax and elytra.

70.—CARENUM TRISTE, MacL., W. *Trans. Ent. Soc.*
N. S. Wales, Vol. 2, page 63.

71.—CARENUM OVIPENNE. n. sp.

Length 6¼ lines.

This belongs to the *C. Bonellii* group. It is entirely black without vestige of striæ on the elytra or any marginal colouring. The anterior puncture on the elytra is further from the shoulder than in *C. Bonellii*, and the body is of a more perfectly oval form.

72.—CARENUM SUBMETALLICUM. n. sp.

Length 7 lines.

This also belongs to the *C. Bonellii* group, and scarcely differs in form from that insect. The thorax is margined with pale green. The elytra are of a dull greenish bronze hue with pale green margin, and are covered with indistinct and interrupted striæ.

73.—CARENUM DEAURATUM, MacL., W. *Trans. Ent. Soc. N. S. Wales, Vol. 1, page 140.*

74.—CARENUM ANGUSTIPENNE. n. sp.

Length 8 lines.

Black, subnitid, with a narrow pale green margin to the

elytra, which are narrower than the thorax and narrowest at the
base. In form this insect resembles *C. quadripunctatum* mihi,
and it belongs to the same group. The antennæ are more
slender and filiform than in the typical *Carenums*.

75.—EUTOMA BIPUNCTATUM, MacL., W. *Carenum bipunctatum, MacL., W., Trans. Ent. Soc. N. S.
Wales, Vol.* 1, *page* 60.

76.—SCARITES CACUS, MacL., W. *Trans. Ent. Soc.
N. S. Wales, Vol.* 1., *page* 67.

77.—EUDEMA ALTERNANS, Casteln. *Not. Aust. Col.*
1867, *page* 60.

78.—EUDEMA AUSTRALE, Dej. *Spec. Col., Vol.* 5, *page*
601.

79.—EUDEMA AZUREUM, Casteln. *Not. Aust. Col.*
1867, *page* 61.

80.—CHLÆNIUS AUSTRALIS, Dej. *Spec. Col., Vol.* 5,
page 650.

81.—CHLÆNIUS MACULIGER, Casteln. *Not. Aust. Col.*
1867, *page* 62.

82.—CHLÆNIUS PEREGRINUS, Laferté. *Ann. Fr.* 1851,
page 247. Chaud. *Bull. Mosc,* 1856, 3, *page*
264.

83.—CHLÆNIUS MARGINATUS, Dej. *Spec. Col., Vol.* 2,
page 305.

84.—OODES AUSTRALIS, Dej. *Spec. Col., Vol.* 5, *page*
671.

85.—PROMECODERUS VIRIDIS. n. sp.

Length 7 lines.

The thorax and elytra of this beautiful species are of a blackish green and very nitid. The head and under surface are of a glossy black, the antennæ, the parts of the mouth and the legs are of a pitchy brown. There are a few impressions on the posterior part of the lateral margins of the elytra.

86.—MEONIS ATER, Casteln.   *Not. Aust. Col.* 1867, *page* 70.

87.—MEONIS OVICOLLIS. n. sp.

Length 3½ lines.

This insect differs from *M. ater*, and *niger*, the two species on which Count Castelnau founded the genus, in being much smaller, in having the mandibles less exserted, and in having the thorax globular and rounded, and in no way cordiform or prolonged at the base. It is of a glossy black with the elytra strongly striated in the middle.

88.—LEIRODIRA LATREILLII, Casteln.   *Not. Aust. Col.* 1867, *page* 72.

89.—PHORTICOSOMUS RUGICEPS. n. sp.

Length 5 lines.

Dark brown with the antennæ and legs of a pitchy red. The head is broad, and much corrugated and wrinkled, the thorax is uneven on the surface, and the elytra are striated with the interstices convex. In all these points it differs from *P. Felix* the type of the genus, which has the head and thorax smooth, and the spaces between the striæ of the elytra rather flat.

90.—LECANOMERUS RUFICEPS. n. sp.

Length 3 lines.

Head red, smooth, and deeply impressed on each side, near the eyes and rather in front of them. Thorax black—with the margin and base slightly reddish,—scarcely broader than the length, rounded at the sides anteriorly, not much narrowed behind, truncate and punctured at the base, deeply impressed near the posterior angles and canaliculate in the middle. Elytra broader than the thorax, moderately convex, strongly striated with the interstices convex; and black, with a broad dull yellow lateral and apical margin. The legs are of a pale yellow.

91.—LECANOMERUS ABERRANS. n. sp.

Length 2 lines.

Black, with the legs yellow, the antennæ and the parts of the mouth red, and with a tinge of the same colour on the margin of the thorax and elytra. Head smooth and lightly impressed on each side in front of the eyes. Thorax rather broader than the length, scarcely narrowed behind, slightly rounded before the middle, truncate at the base, very broadly but not deeply impressed near the posterior angles, and slightly canaliculate in the middle. Elytra strongly striated, the interstices but slightly convex. The anterior tarsi in the male of this species are peculiar, the first joint is very short and not apparently dilated at all, the second is very much dilated, the third is shorter, but quite as much laterally dilated as the second, while the fourth is as small as the first.

92.—HARPALUS INTERSTITIALIS, MacL., W. *Trans. Ent. Soc. N. S. Wales, Vol.* 1, *page* 117.

93.—HARPALUS INFELIX, Casteln. *Not. Aust. Col.* 1867, *page* 106.

94.—HARPALUS PLANIPENNIS. n. sp.

Length 3½ lines.

Head smooth, black, nitid, with a small oval depression on each side near the eyes. Thorax glossy, broader than the length, emarginate in front, truncate behind, slightly rounded on the sides anteriorly, not narrowed behind, canaliculate in the middle, and with a broad and shallow rugose impression at the base between the median line and the posterior angles; these last are rounded. Elytra bronzy, dull, and lightly striated, the interstices flat and broad, with six light punctures at irregular intervals on the third interstice, and a series of deeper ones on the lateral intersticcs. Antennæ, palpi, and legs, with the exception of the upper part of the thighs, of a reddish hue.

95.—HARPALUS GAYNDAHENSIS. n. sp.

Length 3½ lines.

Head and thorax as in the last species. Elytra nitid, with a slight purplish tinge ; the striæ rather deeper than in *planipennis*, with the intersticcs equally flat, and with seven or eight well marked punctures on the third interstice extending from the base to the apex. The thighs are black, the rest of the legs, the antennæ and palpi are reddish.

96.—HARPALUS ANGUSTATUS. n. sp.

Length 3½ lines.

Head and thorax as in the two last species, the latter with the surface less smooth and the posterior impressions less marked. Elytra scarcely broader than the thorax and of a nitid black, with a very slight greenish tinge ; the striæ are well marked as in the last species, and the interstices are flat, with from six to eight well marked punctures on the third, placed at irregular intervals from the base to the apex. Thighs black, the rest of the legs, antennæ and palpi are reddish.

97.—HARPALUS CONVEXIUSCULUS. n. sp.

Length 3 lines.

Glossy black with the legs and palpi red. Head and thorax smooth and convex, the latter with the posterior impressions deep, short, and narrow, and with the lateral margins red and furnished with several setigerous punctures. Elytra convex, very little broader than the thorax, lightly striated, with abbreviated striæ near the suture, and having the interstices broad and nearly quite flat.

98.—HARPALUS ÆNEO-NITENS. n. sp.

Length 3 lines.

Brassy black with a tinge of purple on the elytra. Head

smooth. Thorax rather flat, rounded at the sides, truncate behind, and rugose on the back, especially near the base, with the median line slight and not reaching to the base, and the posterior impressions broad and rather deep. Elytra flat and distinctly but not deeply striated, with the interstices broad and flat, and with five or six punctures on the third interstice, and one on the fifth near the apex. The abbreviated sutural striæ are longer than in *H. Gayndahensis*, the species to which it most approximates. The legs and the parts of the mouth are light red.

99.—HARPALUS ATRO-VIRIDIS. n. sp.

Length 2¾ lines.

More convex than the species last described, but not so much so as *H. convexiusculus;* of a nitid greenish black colour, with yellow legs and palpi. Head broad and smooth. Thorax with a very narrow reddish margin, the median line only marked in the middle, the posterior impressions deeply marked but small, and the sides moderately rounded. The elytra are deeply but narrowly striated, with the interstices broad and flat. There is a small puncture on the third interstice near the apex, and the abbreviated sutural striæ are short, and seem to take their rise from the extremity of the second striæ. The legs are more than usually armed with short spiniform hairs.

100.—STENOLOPHUS POLITUS. n. sp.

Length 2½ lines.

Black and nitid. Head rugosely impressed on each side before the eyes. Thorax nearly square, slightly rounded at the sides which have a narrow red margin, and broadly impressed and slightly rugose near the base on each side of the median line which is very lightly marked. Elytra lightly striated,—the interstices flat and broad,—with an obscure spot on the shoulder, a narrow lateral margin enlarged towards the apex, and a narrow edging on the apical half of the suture, of a yellowish red. The legs are of a light red with the apices of the thighs, tibiæ, and tarsi, dark brown. The antennæ and palpi are of a brownish red.

101.—ACUPALPUS MASTERSII. n. sp.

Length 2 lines.

Black, nitid, rather convex. Head with a short curved impressed line on each side from the eye to the transverse line of the clypeus. Thorax broader than the length, and rounded on the sides and at the posterior angles. The basal impressions are scarcely traceable, and the median line is but lightly marked. The elytra are strongly striated and have the interstices slightly convex. The legs, antennæ, and palpi are of a reddish yellow.

102.—ACUPALPUS ANGULATUS. n. sp.

Length 2 lines.

Black, nitid, and moderately convex. Head as in the last species. Thorax narrower, longer than the breadth, rounded at the sides anteriorly, and slightly narrowed behind, with the posterior angles almost square and broadly margined, and the median line and basal impressions well marked. Elytra considerably broader than the thorax, with the striæ not quite so deep, and the interstices rather more flat than in the last species. The legs and antennæ are of a pale red.

CYCLOTHORAX. n. gen.

Mentum deeply emarginate with a large median tooth. Labium rather long, obtuse at the apex. Palpi short, somewhat filiform, the tip of the maxillary rather obtuse. Labrum square, entire. Mandibles short, strong, and slightly arcuated with a small tooth in the centre of the right mandible. Antennæ of moderate length ; 1st joint long and thicker than the others ; 2nd small ; 3rd and 6th longer than the others ; the remainder equal. Thorax convex, transversal and rounded at the sides and base. Elytra broader than the thorax, slightly convex, and short. Legs moderately strong, the anterior tarsi slightly dilated in the male, the intermediate still less so ; the two first joints of all the tarsi longest.

I am not at all sure of the position of this genus, the dilatation of the tibiæ in the male is so slight as to be in most instances unnoticeable.

103.—CYCLOTHORAX PUNCTIPENNIS. n. sp.

Length 2¾ lines.

Black, nitid. Head smooth with a deep line on each side near the eyes. Thorax emarginate in front, much rounded at the sides and base, with a narrow red reflexed margin ; the basal portion depressed and punctured, and the remainder smooth and without median line. Elytra each with six light striæ marked with well defined punctures, which become less distinct towards the apex, and in the sixth striæ do not extend beyond the middle ; there is also a punctured sublateral stria. A very narrow reddish border is traceable along the sides and hinder portion of the suture. The legs, antennæ, and palpi, are of a pale red.

104.—ABACETUS ATER. n. sp.

Length 2¾ lines.

Black, nitid. Head small with a short deep curved impression on each side in front. Thorax much broader than the head, and a little broader than the length, emarginate anteriorly, much rounded on the sides, and truncate at the base. The median line is distinct, and takes its rise at some distance from the apex, and the two basal impressions are deep, long, and linear. The lateral margin is small, reflexed, and of a reddish hue. Elytra scarcely broader than the thorax at its broadest part, and deeply striated, with small punctures visible in the striæ, and with the interstices broad and convex. Legs, antennæ, and palpi pitchy red.

105.—ABACETUS ANGUSTIOR. n. sp.

Length 2¾ lines.

This species nearly resembles the last ; it differs in the thorax, which is less emarginate in front, less bulged out and rounded on the sides, and which is rather longer than the width. The elytra differ in having the striæ smaller, and the interstices much more flat. The legs and palpi also are of a lighter colour.

106.—AMBLYTELUS AMPLIPENNIS.   n. sp.

Length 5 lines.

Reddish brown, nitid.   Head flat, with two broad impressions between the eyes.   Thorax scarcely broader than the head, nearly square, the anterior angles obtuse, the posterior acute, the median line distinctly marked, the basal impressions broad and very deep, and the surface marked with transverse striolæ, which become more profound towards the base.   Elytra narrowest at the shoulders, widest near the apex, which is broadly rounded ; marked with light faintly punctured striæ, and with the interstices broad and nearly flat.   The under surface and legs are of a lighter hue.

This insect ought perhaps to constitute another genus.   The median tooth of the mentum, though large, is not nearly so much so as that of *Amblytelus* is said to be, and the form of the thorax is also very different, but the general resemblance and evident affinity are very great.

107.—AMBLYTELUS MINUTUS.   n. sp.

Length 2½ lines.

In colouring and marking almost identical with the last species, it differs in being very minute, in having the facial impressions more profound and linear, and in having the thorax with a larger and more reflexed lateral margin, and less profound basal impressions.

108.—LESTICUS CHLORONOTUS, Chaud.   *Ann. de la Soc. Ent. Belg. T.* XI.

109.—TIBARISUS ATER.   n. sp.

Length 7½ lines.

Black, nitid.   Head smooth with two longitudinal impressions on each side, one extending from the eyes to the root of the mandibles, the other,—a broader impression,—situated between and rather in front of the eyes.   Thorax square, truncate behind and nearly so in front, slightly rounded at the sides and narrowly margined laterally, with the median line distinct, and with two basal impressions on each side, the outer one short and curved, the inner long and deep. Elytra very profoundly striated with the interstices smooth and convex. Tarsi clothed with red hair.

This species differs from *T. melas*, Casteln., in being much smaller, and in having no vestige of an abbreviated stria near the scutellum, whereas, in *T. melas* there is a small impression on the first interstice, emanating from the basal end of the second stria.

110.—TIBARISUS NIGER. n. sp.

Length 5 lines.

This species differs from the last in its much smaller size, in its palpi being less truncate, in the basal impressions of the thorax being broader, and in having a distinct abbreviated stria near the scutellum.

111.—OMALOSOMA HERCULES, Casteln. *Not. Aust. Col.* 1867, *page* 119.

112.—OMALOSOMA MARGINIFERUM, Chaud. *Bull. Mosc.* III, *page* 68.

113.—OMALOSOMA CORDATUM, Chaud. *Bull. Mosc.* III, *page* 69.

114.—OMALOSOMA CUNNINGHAMII, Casteln. *Not. Aust. Col.* 1867, *page* 120.

115.—NOTONOMUS PURPUREIPENNIS. n. sp.

Length $7\frac{1}{2}$ lines.

Head and thorax black with a greenish tint, the latter very little longer than the breadth, very slightly rounded at the sides, of the same width at the anterior and posterior angles, emarginate at the base, and deeply impressed on the median line, with the basal impressions long and linear. Elytra purple with a green border, and strongly striated, with the interstices convex, the third marked with two large punctures, one near the centre the other between that and the apex. Legs ciliated with reddish hair.

This may be the *Feronia impressicollis* of Castelnau, if so, he has made a mistake in describing the posterior margin of the thorax, as being cut convexly in the middle.

116.—NOTONOMUS VIOLACEOMARGINATUS. n. sp.

Length 7½ lines.

This species is exactly like the last excepting that the elytra have a brilliant red violet margin, are opaque, and have the interstices between the striæ almost flat.

117.—NOTONOMUS CYANEOCINCTUS. n. sp.

Length 6 lines.

Black, nitid. Thorax longer than the breadth, scarcely narrowed behind, and slightly rounded at the sides, with the median line and basal impressions deeply marked ; the latter are broader and shorter than in the two preceding species. Elytra deeply striated, with the interstices convex, the third interstice marked with two large punctures, one a little behind the middle, the other between that and the apex ; the lateral margin is of a brilliant blue changing in some lights to green.

118.—NOTONOMUS VIRIDICINCTUS. n. sp.

Length 5½ lines.

Head and thorax black and nitid with a purplish gloss, the latter longer than the breadth, rounded on the sides anteriorly, a little narrowed behind, and emarginate at the base, with the median line broadly marked, and the basal impressions long and deep. The elytra are of a rather dull purple with a bright green margin, and strongly striated, with the interstices but slightly convex, and with two large punctures on the third interstice, one behind the middle, the other near the apex. Legs clothed with reddish hair.

119.—NOTONOMUS ANGUSTIPENNIS. n. sp.

Length 5 lines.

Differs from the last species in being less brilliant throughout, in having the thorax less rounded laterally, in having the elytra scarcely broader than the thorax, in being destitute of all purple gloss except on the elytra, where it is very slight, and in having the abbreviate stria near the scutellum shorter and more pointed. In other respects the two species are exactly alike.

120.—STEROPUS CYANEOCINCTUS, Chaud. *Bull. Mosc.*, 1865, III, *page* 97.

121.—OMASEUS MASTERSII. n. sp.

Length 6½ lines.

Black, very nitid. Head smooth in the middle, with two small elongate impressions on each side, the external one linear and extending from the eyes to the base of the mandibles, the other impression rather broader. Thorax almost square, rounded behind the anterior angles, truncate at the base, and very slightly narrowed at the posterior angles, with the median line lightly marked and with two deep basal impressions on each side; the outer one close to the posterior angle and very short, the other moderately long and very broad. Elytra of a brilliant iridescent hue, with four deep and distinct striæ on each side of the suture, and a lateral stria, perfect, with the fifth stria from the suture scarcely traceable towards the apex, and the sixth and seventh only traceable at the apex. The first four interstices are convex.

122—CHLÆNIOIDEUS PLANIPENNIS. n. sp.

Length 8½ lines.

Black, nitid. Head deeply but shortly bisulcated in front. Thorax rather broader than the length, slightly rounded at the sides anteriorly, scarcely narrowed behind, broader at the posterior angles than at the anterior, and truncate at the base, with the median line very lightly marked, the inner basal impressions broad and deep, and the outer almost circular. Elytra broad and lightly but distinctly striated, with the abbreviated stria near the scutellum two lines long. The interstices are broad and very flat.

This species differs from *C. prolixus* Erichs., upon which the genus is founded by Baron de Chaudoir, in several particulars, the most evident being the much finer striation of the elytra and greater flatness of the interstices.

123.—PŒCILUS SUBIRIDESCENS. n. sp.

Length 4½ lines.

Black, nitid. Head with two short impressions in front. Thorax longer than the breadth, rounded at the sides, rather narrowed behind, truncate at the base, and with the posterior angles obtuse and rounded, the median line lightly marked, and the basal impressions deep and rather long. Elytra iridescent and strongly striated, with no abbreviated sutural striæ, and with the interstices convex. The legs and antennæ are of a reddish brown.

124.—PŒCILUS ATRONITENS. n. sp.

Length 4 lines.

In form and general appearance very much resembling *P. subiridescens*. It differs in having the frontal impressions smaller and more lightly marked, and in having the spaces between the striæ on the elytra more flat. It is also a much smaller insect, and with only a trace of iridescence on the elytra.

125.—ARGUTOR FOVEIPENNIS. n. sp.

Length 3½ lines.

Black, nitid. Head with the frontal impressions long and well marked. Thorax transversal, rounded behind the anterior angles and square at the base with the median line well marked, the inner basal impression broad and deep, and the outer very small. Elytra duller than the thorax,—of a brownish tinge, with an indistinct dull red margin,—and striated, with the abbreviated stria long and joining the first stria from the suture, and with the interstices flat ; the third with three shallow foveæ, the first above the middle on the outside of the interstice, the second about the middle on the interstice, and the third between that and the apex on the inside of the interstice. The legs and antennæ are of a pitchy red.

126.—ARGUTOR NITIDIPENNIS. n. sp.

Length 3 lines.

Black, nitid. Head with the frontal impressions long. Thorax of the same form as in *A. foveipennis*, but with the basal impressions broader, shallower, and somewhat rugose and punctured. Elytra brilliant and striated with the abbreviated basal stria long and joining the first stria, and with three punctures on the third interstice, the upper one not far from the base, and on the outer side of the interstice. The legs and antennæ are red.

127.—ARGUTOR OODIFORMIS. n. sp.

Length 3½ lines.

Black, nitid. Head as in the last, but with the frontal impressions more rugose. Thorax rather more rounded at the anterior angles than in the two preceding species, and having much the appearance of an *Oodes*. Elytra with a slight bronzy tinge, and strongly striated, with the interstices convex, the third having three punctures as in the last described species; the abbreviated stria near the scutellum does not join the 1st stria. The tarsi and antennæ are of a brownish red.

128.—PLATYNUS NITIDIPENNIS. n. sp.

Length 4 lines.

Black, very brilliant, with the elytra iridescent, the legs, antennæ, and palpi red, and the labrum reddish brown. Head broadly impressed on each side anteriorly. Thorax nearly square, very narrowly margined, a little rounded on the sides anteriorly, very slightly narrowed behind, and truncate at the base, with the posterior angles obtuse, the median line distinct and the basal impressions long, narrow, and deep. The elytra are a little broader than the thorax and more than twice the length, and

H

strongly striated, with the interstices convex, and without an abbreviated stria near the scutellum.

This may not be properly one of the *Anchomenidæ*, it much resembles some of the *Feronidæ*. The palpi, however, and the elongate antennæ and legs induce me to place it in this genus.

129.—PLATYNUS PLANIPENNIS. n. sp.

Length 5¼ lines.

Of a brilliant bronzy black above, and of a pitchy black below, with the margin of the thorax, and the extreme lateral edge of the elytra red. Head smooth with the frontal impressions not deep. Thorax nearly square, rounded on the sides, a little narrowed posteriorly, truncate at the base, with the posterior angles obtuse and the margin reflexed; the median line is distinctly marked from the anterior transverse impression to the posterior, and the basal impression on each side is very broad and shallow. The elytra are finely striated, with perfectly flat interstices; the third interstice has on or near it three large punctures,—the first a little distance from the base in the third stria, the second near the middle in the second stria, and the third nearer the apex in the same stria; the abbreviated stria near the scutellum is long.

130.—PLATYNUS MARGINICOLLIS. n. sp

Length 5 lines.

Head more deeply marked than in the last described species. Thorax the same. Elytra narrower and of a greenish hue, with the reddish border larger, and the striæ closely punctured; the interstices between the striæ also appear less flat and broad, and have a shagreen appearance. The legs are yellow. In other respects like *P. planipennis*.

131.—DICROCHILE GORYI, Boisd. *Voy. Astrol. Col.*, *page 32.*

SIAGONYX. n. gen.

Mentum rather large without median tooth, and with the lateral lobes acutely pointed at the inner portion of the apex.

Labium obtuse and corneous at the apex. Palpi long and slender, the second joint of the maxillary longest, the terminal joint of both thickened, and obtuse or truncate at the tip. Mandibles strong and terminating in two very acute arcuated teeth. Labrum short, broad and entire or very slightly bi-emarginated. Antennæ long and slender, the first article long and thicker than the others, the second short. Head small, triangular, and slightly narrowed behind. Thorax elongate and narrowed behind with a broad recurved margin. Elytra very slightly convex and of an elongated oval form. Legs long and slender, the first joint of the tarsi very long, fourth short and entire, the first three joints of the anterior tarsi of the male dilated.

This genus resembles *Lestignathus*, Erichs., it differs chiefly from it in the very different labium, labrum, and mandibles.

132.—SIAGONYX AMPLIPENNIS. n. sp.

Length 7 lines.

Black, moderately nitid. Head rather rugosely impressed on each side in front. Thorax rounded at the sides, narrowed behind, reflexed broadly at the posterior angles, rugose at the base with the median line deeply marked, and the basal impressions broad. Elytra much broader than the thorax, broadest behind the middle, and strongly striated, with the interstices convex. Tarsi and antennæ reddish brown.

133.—SIAGONYX MASTERSII. n. sp.

Length 6 lines.

This species differs from the preceding in having the head smoother, the thorax more narrow, and in being altogether a smaller and more elongate looking insect. In other respects the two species exactly resemble one another.

134.—TRECHUS ATRICEPS. n. sp.

Length 1¼ lines.

Head black, smooth, and deeply impressed on each side in front, with the clypeus, labrum, palpi, and two basal joints of the antennæ, yellow. Thorax red, almost square, rounded at the sides anteriorly, narrowed behind, and truncate at the base, with the median line distinct and the basal impression large, shallow, and punctured. Elytra dark brown and striated, with the margin and suture red. Legs of a pale red.

135.—TRECHUS RUFILABRIS. n. sp.

Length 1½ lines.

Head black, smooth and very lightly impressed in front, with the clypeus, labrum, palpi, and antennæ of a yellowish red. Thorax yellowish red, very little wider than the head, rounded slightly on the sides, narrowed behind and rounded at the posterior angles, with the median line very lightly marked, and the basal impressions very shallow and punctured. Elytra flat, smooth, obsoletely striated, shorter than the body, and of a yellowish red colour, with a large dark brown spot covering nearly all the apical half of each elytron. The under side of the body and legs are of a pale red.

136.—TRECHUS CONCOLOR. n. sp.

Length 1¾ lines.

Entirely of a yellowish red colour, excepting the antennæ from the third joint to the apex which are of a darker hue, and a slight dash of brown on the forehead. Thorax almost square, rounded a little on the sides anteriorly, and slightly narrowed behind, with the median line slightly marked, and the basal portion transversely impressed and punctured. Elytra strongly striated, the interstices slightly convex.

137.—TRECHUS ATER. n. sp.

Length 1¾ lines.

Black, rather dull. Head deeply impressed on each side. Thorax square and very little narrowed behind, with the median line distinct, and the basal impressions very broad, very shallow, and densely punctured. Elytra strongly striated, with a narrow reddish margin, and a slight tinge of the same colour along the apical half of the suture. The legs, palpi, and first joint of the antennæ are red, the rest of the antennæ brown and pilose.

The following seventeen species have subulate palpi, and though comprising many different forms, may all be referred to the very comprehensive genus *Bembidium*. I have not attempted to divide them into subgenera.

138.—BEMBIDIUM BISTRIATUM. n. sp.

Length 1 line.

Black, convex, very nitid. Head deeply impressed on each side in front. Thorax somewhat globular, transverse, rounded on the sides, a little lobed at the base and slightly impressed on the median line. Elytra ovate, with a deep stria on each side of the suture, and two lateral striæ, the inner one punctured behind ; there are also two large dark red round spots on each elytron, one near the shoulder, the other near the apex. The legs, antennæ, and palpi are yellow, the latter have the penultimate joint much swelled.

139.—BEMBIDIUM STRIOLATUM. n. sp.

Length 1 line.

Black with a brilliant bronzy lustre. Head and thorax like those of *B. bistriatum*, the latter not so globular, nor so rounded on the sides, and with the base more truncate and the posterior angles square. Elytra ovate and striated, the striæ broad, rather shallow, and not extending to the base or apex. There are two round yellow spots on each elytron, placed much as in the last species, but much smaller. The legs and basal half of the antennæ are of a reddish yellow, the rest of the antennæ is of a dark colour.

140.—BEMBIDIUM CONVEXUM. n. sp.

Length 1 line.

This species only differs from *B. bistriatum* in being more convex, and in having the thorax more rounded and narrowed posteriorly, and in having the impressions or foveæ on the basal margin of the elytra more profound.

141.—BEMBIDIUM BIPUSTULATUM.   n. sp.

Length 1¼ lines.

Black, subnitid.   Head smooth and longitudinally impressed
on each side between the eyes.   Thorax rather flat, much broader
than the length, slightly rounded on the sides, not narrowed
behind, and truncate at the base, with the median line scarcely
visible and the basal impressions very short.   Elytra broad,
striated and of a bronzy hue, with an indistinct red spot near the
apex on the outside of each elytron.   Legs, palpi, and antennæ
yellow.

142.—BEMBIDIUM PUNCTIPENNE.   n. sp.

Length 1 line.

Entirely of a glossy red.   Head large and flat, with the impressions between the eyes long and deep.   Thorax transversal,
convex, much rounded on the sides and narrowed at the base,
with the median line lightly marked.   Elytra moderately convex,
and marked with regular rows of distinct punctures which disappear towards the apex.   The suture and lateral margins of the
elytra are of a rather darker hue than the rest of the body.

143.—BEMBIDIUM ATRICEPS.   n. sp.

Length 1 line.

Head black, and largely impressed on each side between the
eyes.   Thorax yellowish red, rather flat, transversal, slightly
rounded on the sides anteriorly, and with the posterior angles
square and broadly reflexed.   Elytra somewhat flat, strongly
striated, punctured in the striæ, and of a reddish colour, with the
suture, apex, and a broad median fascia, black.   Legs, antennæ,
and palpi yellow.

144.—BEMBIDIUM TRANSVERSICOLLE.   n. sp.

Length 1⅙ lines.

Of a pale red, slightly clouded with brown on the head and
elytra.   Thorax transversal, convex, rounded on the sides in front,
square and acute at the posterior angles, and transversely de-

pressed at the base, with the median line distinct. The elytra are marked with indistinct striæ, those next the suture being the most distinct.

145.—BEMBIDIUM SEXSTRIATUM. n. sp.

Length 1½ lines.

Upper surface of a pitchy red clouded in some parts with brown. Head with the frontal impressions broad and long. Thorax little convex, transversal, rounded on the sides, and but very little narrowed towards the posterior angles, which are square and broadly reflexed; the median line is well marked, and the centre of the base is transversely depressed and rugose. The elytra have three distinct broad striæ on each side of the suture. beyond these other striæ may be traced, but they are very indistinct, there is also a large puncture about the line of the fourth stria, at a short distance from the base. The legs and first joint of the antennæ are yellow.

146.—BEMBIDIUM OVATUM. n. sp.

Length scarcely 1 line.

Entirely of a pale testaceous colour, convex, and broadly ovate. Head with the frontal impressions long and linear. Thorax transversal, rounded on the sides in front, not narrowed behind, square and slightly reflexed at the posterior angles, and truncate at the base, with a very lightly marked median line and with the transversal depression of the base very small. Elytra broadly ovate with a distinct stria on each side of the suture, and a deep fovei-form impression on the basal margin on each side of the scutellum.

147.—BEMBIDIUM BIFOVEATUM. n. sp.

Length ¾ of a line.

This species only differs from the last in being smaller, less broad, in having no trace of striæ on the elytra beyond the distinct one on each side of the suture, and in having a deep foveiform impression at the base of the thorax on each side of the centre, which extends to the elytra, and is in fact common to both.

148.—BEMBIDIUM BRUNNIPENNE. n. sp.

Length 1 line.

Head and thorax reddish brown, the former finely punctured with the frontal impressions lightly marked. Thorax slightly convex, transversal, punctured, rounded on the sides in front, and not narrowed behind, with the posterior angles square, the transversely depressed portion of the base rather large, the median line deeply marked, and the basal impressions short. The elytra are of a dark brown, rather flat, finely punctured all over and striated. The legs, antennæ, and palpi are pale red.

Mr. Masters informs me that he found this species on stony ground, and on the summit of a high hill remote from water, a habitat very different from that of this family of insects generally.

149.—BEMBIDIUM RUBICUNDUM. n. sp.

Length ¾ of a line.

Red, subnitid. Head with the frontal impressions broad but not very deep. Thorax transversal, rounded on the sides, and slightly narrowed at the posterior angles which are square, with the median line very lightly marked, the basal impressions short, and the transversely depressed portion of the base strongly marked. The elytra are rather depressed, and have about six rows of distinct punctures on each side of the suture, extending from the base to beyond the middle.

This species very much resembles *B. punctipenne*, it is however a much smaller insect, and the shape of the thorax is different being square behind in the present species, while in the other it is very much rounded.

150.—BEMBIDIUM SUBVIRIDE. n. sp.

Length 1½ lines.

Greenish black, with a metallic reflection. Head broad, smooth, and lightly impressed on each side. Thorax scarcely broader than the head, transversal, rounded on the sides, and not narrower behind than in front, with the posterior angles acute and slightly reflexed, and with the median line distinctly marked. Elytra rather flat, finely striated, the striæ finely punctured, and the interstices smooth, broad, and flat, with the apex and a lateral spot near the apex, orange. The legs, antennæ, and palpi, are of a pitchy red.

151.—BEMBIDIUM AMPLIPENNE. n. sp.

Length 1½ lines.

Entirely of a testaceous red colour. Head with the frontal impressions short and linear. Thorax broader than the head, transversal, very slightly rounded on the sides, and truncate at the base with the posterior angles square, the median line distinct, the basal transverse depressions very deeply marked, and the basal impressions short. Elytra broad, moderately convex and striated, the striæ large and deep, and the interstices convex. The legs are yellow.

152.—BEMBIDIUM GAGATINUM. n. sp.

Length 1½ lines.

Black, subnitid, and of an oblong form. Head broad, convex, and smooth, without any impression behind the clypeus. Thorax a little wider than the head, slightly rounded on the sides, not narrowed behind, with the posterior angles very obtuse, the basal impressions round, the median line scarcely traceable, and the basal portion of the whole thorax lightly rugose and punctured. The elytra are a little wider than the thorax, parallel sided, very finely striated, the striæ full of small punctures, and most distinct towards the apex, at the base scarcely traceable. The legs are of a pitchy hue.

153.—BEMBIDIUM FLAVIPES. n. sp.

Length 1½ lines.

Head exactly as in the last species. Thorax also exactly as in the last species, excepting in being more narrowed behind, and in having the median line distinct. Elytra large, rather depressed, widening from the base to near the apex, and faintly striated and punctured. The legs are yellow.

154.—BEMBIDIUM BIPARTITUM. n. sp.

Length 1½ lines.

Antennæ nearly as long as the body. Labrum deeply emarginated. The terminal subulate joint of the palpi long. Head red, long, and very deeply canaliculated on each side from behind the eyes to the labrum. Thorax also red, transversal, convex, much rounded at the anterior angles, acutely pointed and slightly reflexed at the posterior angles, and slightly prolonged into a lobe at the base, with the median line deeply marked and the anterior transverse line making a deep emarginate impression in front. The body is pedunculated as in *Scarites*. Elytra jet black, very nitid, of an oblong form and rather depressed, with a deep stria on each side of the suture, and two large punctures, one near the base, the other below the middle of each elytron. The legs are strong and of a pale red.

This species ought no doubt to form a new genus. I have not, however, been able to get a specimen for dissection, and cannot consequently undertake to give the generic characters correctly.

The three following species were accidentally omitted in their proper places in this list.

155 —HELLUOSOMA ATRUM, Casteln. *Not. Aust. Col.* 1867, *page* 21.

156.—BADISTER ANCHOMENOIDES. n. sp.

Length 3 lines.

Black, nitid, with a slight iridescence on the elytra. Head smooth, not deeply impressed on each side in front of the eyes, with the labrum thick, membranaceous looking, and triangularly emarginated. Thorax square and obtusely angled with the basal impressions short but deep, and the median line distinct. Elytra striated with the interstices flat. Legs, antennæ, and margin of thorax and elytra yellow.

157.—PHYSOLŒSTHUS GRANDIPALPIS. n. sp.

Length 2¾ lines.

Brownish black, subnitid and of rather flattened form. Head as in the last species, with the last joint of the palpi large and obliquely truncated. Thorax rather transversal, with the anterior angles advanced, the posterior square and slightly reflexed, the median line deep, and the basal impressions broad and long. Elytra of an elongated ovate form, strongly striated and with the interstices convex. Legs, antennæ, and palpi reddish.

This completes the *Carabidæ* in the Gayndah collection, with the exception of five species of *Olivina*. These I cannot venture to name as I have been unable to procure the works of M. Putzeys, who has paid special attention to this group, and who has described, I believe, a large number of Australian species.

DYTISCIDÆ.

158.—HYDROPORUS BIFASCIATUS. n. sp.

Length 2 lines.

Ovate, subconvex, covered with a close puncturation, and of a testaceous red colour. Head bordered in front, and with a very short oblique lightly impressed line at the inner and anterior angle of the eyes. Thorax broad, short, narrowly margined on the sides, and broader behind than in front, with the anterior angles advanced, the posterior subacute, the base slightly bisinuate on each side, and a large patch of a dark brown colour on each side of the central lobe. Elytra broader than the thorax, rounded at the humeral angles and on the sides, and narrowed and rounded at the apex, with a very zigzag black fascia about the middle, extending from the suture to near the sides, and another of the same hue, and rather broader near the apex, extending from the sides almost to the suture, the two fasciæ being joined near the suture, and nearly joined about the middle of the width. On each elytron may be traced an obsolete stria a little way from the suture, and also a number of small round obsolete looking depressions. The tarsi of the male are black and much dilated, the third joint being the largest.

159.—HYDROPORUS GIGAS, Bohem. *Res. Eugen.* 1858, *page* 18.

160.—HYDROPORUS FOVEICEPS. n. sp.

Length 3 lines.

Ovate, flat, very finely punctured, and of a testaceous red colour. Head with a small round fovea on each side between and in front of the eyes. Thorax broad, short, narrowly margined on the sides, and broader behind than in front, with the anterior angles advanced, the posterior angles square, the base very slightly sinuated and the apical and basal margin tinged with brown. Elytra of the same width as the thorax at the base, gradually widening to the middle and rather narrowly rounded at the apex, with two obsolete striæ on each, one close to the suture, the other nearer the middle. In colour the elytra are much paler than the rest of the insect and of a more lurid hue, with the suture and two very large patches nearly covering the apical two thirds, of a dark brown. The legs and under side of the body are red. The last joint of the four anterior tarsi is long and slight.

161.—HYDROPORUS BRUNNIPENNIS. n. sp.

Length 2½ lines.

Of the same form and sculpture as the last described species. The colour is pitchy red, with the thorax margined in front and at the base with brown, and with the elytra of a dark brown with reddish lateral margin.

162.—HYDROPORUS FOSSULIPENNIS. n. sp.

Length 2 lines.

Ovate, convex, very finely and closely punctured, and of a testaceous red colour. Head margined in front, and with a large shallow depression on each side. Thorax broad, short, margined at the sides, and broader behind than in front, with the anterior angles advanced, the posterior angles sub-acute, and the base brown, slightly bisinuated on each side of the centre, and marked with short rough "striolæ." Elytra broadly ovate, with the suture and two vittæ,—the first near the suture, the other about an equal distance outside of the first, neither reaching the base, but joining together towards the apex, and forming a very large patch,—of a deep black. A light short stria extends from the base to about the middle in the line of the vitta nearest the suture, and in the middle of the second vitta, there is a deep, short, longitudinal impression.

163.—HYDROPORUS NEBULOSUS. n. sp.

Length 1½ lines.

Much smaller than the last insect, less convex, and more coarsely punctured. The lateral depressions on the head are very shallow. The thorax, which like the head, is of a dull red, is narrowly margined with brown at the base. The elytra are rather pointed at the apex, and have at their base traces of striæ, which are, with the exception of the sutural one, very short; their colour is dark brown, with a sub-basal fascia, a lateral vitta, the apex, a sub-apical narrow zigzag fascia, and a central spot, of a dull red. The legs are pale red, with a narrow band of black on the extremity of each joint of the posterior tarsi.

164.—HYDROPORUS MASTERSII. n. sp.

Length 1¼ lines.

Of an elongated oval form, and pale red colour. Head finely punctured with a well marked fovea near the anterior angles. Thorax also finely punctured, broad, short, rounded slightly at the sides, and broader behind than in front, with the anterior angles advanced, the posterior angles square, the base very slightly bisinuated, the central lobe rounded, a deep basal striola curved inward on each side intermediate between the central lobe and the posterior angles, and the basal margin between these striolæ of a dark brown. Elytra finely punctured, with a distinct stria on each side of the suture, and the thoracic striola continued on them in a straight line to nearly twice the length of that on the thorax. The colour of the elytra is a deep dark brown, with a very broad zigzag fascia enclosing an elongated brown spot, a spot behind the middle, the apex, and the lateral margins, of a pale red.

165.—HYDROPORUS LURIDUS. n. sp.

Length 1¼ lines.

Only differs from the last in being entirely of a lurid hue, with the base of the thorax brown, and some indistinct patches of the same colour on the elytra, these last are proportionately broader and are entirely without the subsutural stria.

166.—HYDROPORUS BASALIS. n. sp.

Length ⅔ of a line.

Ovate, rather depressed, very finely punctured. Head dark brown, nitid, with scarcely a trace of lateral depressions. Thorax of the same form as the two last described species, of a yellow colour, with a narrow apical and broad basal margin of a deep brown, and with the basal striolæ the same as in *H. Mastersii* and *luridus*. Elytra deep brown, with the sides and two somewhat indistinct vittæ of variable length, but not reaching to the base or apex, of a yellowish hue, and with the subsutural stria distinct and the thoracic striola continued to the middle of each elytron.

167.—HYDROPORUS POLITUS. n. sp.

Length ¾ of a line.

This may probably not belong to the present genus. It is of an elongate convex form almost acuminated at the apex of the elytra and very nitid on the entire surface. The head and thorax are of a dark red; the elytra are of a pitchy black; the last six or seven joints of the antennæ seem moniliform.

NECTEROSOMA. n. gen.

Head, antennæ, and parts of the mouth as in *Hydroporus*. Body short, thick, ovate and slightly depressed. Anterior thighs and tibiæ long and stout, the latter in the male largely emarginated in the middle of the internal surface. The four anterior tarsi are five jointed, the first three large, more especially in the male, the two last long and slender, the last very long.

There are several species which will come under this genus, which have been described by the late Rev. Hamlet Clark as *Hydropori*, such as *H. penicillatus*, *Wollastonii, dispar., &c.*

The shape, however, of the anterior legs and the five jointed tarsi, necessitate their removal into another *genus*.

168.—NECTEROSOMA VITTIPENNE. n. sp.

Length 2 lines.

Ovate, broad, subconvex and of a yellow colour. Head with a shallow slightly elongated fovea on each side in front. Thorax broad, short, and bisinuate at the base, with the basal lobe rounded, the posterior angles acute, the anterior angles advanced, the apical border brown, the basal border also brown and enlarged into two spots about midway between the centre and the posterior angles, and with a small striola at the base on the outer side of these brown spots, and a series of smaller striolæ along the whole basal border. Elytra dark brown, with five yellow vittæ on each side of the suture, extending from the base to the apex, with large spots of the same colour on the sides, and with one or two oblique striolæ on the scutellar region, giving the appearance of a large scutellum.

169.—NECTEROSOMA FLAVICOLLE. n. sp.

Length 1¾ lines.

This species differs from the last in having the frontal depression more round and very shallow, in having the thorax entirely yellow, with a depression at the base equidistant from the centre and the posterior angles, in which are three or four small but distinct striolæ, and in having the elytra without striolæ on the scutellum region, and of a yellow colour, with a series of six vittæ, and some lateral spots, brown, the second vitta from the suture being abbreviated towards the base and apex. The elytra also are quite acuminated at the apex, much more so than in *N. vittipenne.*

170.—BATRACHOMATUS WINGII, Clark.   *Journ. of Ent.*
Vol. 2, *page* 15.

171.—COLYMBETES AUSTRALIS, Aubè.   *Spec. gen., page*
236.

172.—AGABUS MASTERSII.   n. sp.

Length 3 lines.

Ovate, broad, black, nitid, subconvex, and very finely and closely punctured.   Head with two very short transverse striolæ on each side between the eyes, with a large yellow spot in the middle, and two smaller and darker ones on the vertex.   Thorax very broad and short, somewhat rounded at the base and very slightly bisinuated, and with the anterior angles much advanced, and of a red colour.   Elytra round and rather depressed behind, with a thin yellow vitta extending from the middle to the apex, near the lateral margin, and with one or two minute spots of the same colour inside, and near the commencement of the vitta.   Under side of a nitid pitchy black, with a yellow spot on the sides of each abdominal segment.   The legs are of a pitchy red.

Though much resembling *A. spilopterus* Germ. from South Australia, this species is quite distinct.

173.—COPELATUS AUSTRALIÆ, Clark.   *Journ. of Ent.*
Vol. 2, *page* 20.

174.—COPELATUS IRREGULARIS.   Clark's M.SS. n. sp.

Length 2¾ lines.

Black, depressed, subnitid.   Head with a small subtransversal distinct fovea on each side between the eyes, with a less distinct one behind it, and with all the middle portion from the labrum to the vertex of a dull red.   Thorax broad, short, almost truncate at the base, covered though not closely with short longitudinal striolæ, and with a broad red lateral margin.   Elytra marked with twelve deep striæ on each side of the suture, and with the base, the apex, and a small indistinct central spot of a yellowish red.   The under side is of a pitchy red and very much acuducted.

I have specimens of this insect in my Cabinet from Lizard Island. The name attached to it must have been sent to me by the late Rev. Hamlet Clark, but I cannot find that it has ever been described before under that or any other name.

175.—COPELATUS ELONGATULUS. n. sp.

Length 1¾ lines.

Elongate-ovate, subconvex. Head dark red, with two short fine transverse striolæ on each side between the eyes. Thorax of a brownish red, and nearly truncate at the base, with the median line distinct in the centre only, and with two broad rather wrinkled depressions at the base. Elytra of a cloudy reddish brown, with the base margined with yellowish red, and with several rows of small distinct punctures not very regularly placed, and extending from the base to the apex on each elytron.

176.—CYBISTER GAYNDAHENSIS. n. sp.

Length 12 lines. greatest width of elytra 6 lines.

Of a more elongate form than *C. scutellaris*. Head with a shallow fovea on each side between the eyes, and with a broad yellow border in front. Thorax shaped as in *C. scutellaris* with a transversal anterior impression consisting of small punctures, a median line traceable from the middle to the base, a few very fine oblique striolæ passing from it to the base, and with a broad yellow lateral margin. Scutellum with a slight transverse depression in the middle. Elytra of the same olive hue as in the other species of the genus, with a broad border of yellow extending from the basal angles to the apex near which it is angularly widened, and with two lines of distant punctures on each elytron, much more remote from one another and less distinct than in *C. scutellaris*. The underside of the body is of a pitchy black, with the anterior and intermediate thighs and a spot on each side of the abdominal segments, yellow.

177.—EUNECTES PUNCTIPENNIS. n. sp.

Length 6½ lines.

Oblong-ovate and of a pale luteous colour. Head with a very

I

small fovea on each side between the eyes, and with a short transverse slightly incurved black band in the centre, and two round spots of the same colour on the summit of the vertex. Thorax broad, and widening gradually in an almost straight line from the anterior angles to the posterior which are acute, with the apex somewhat lobed in the centre, and of a deep black along the line of the apical transverse line, and with the base nearly truncate and marked with two transverse patches of a dark colour meeting in the centre and not extending to the posterior angles. Elytra closely covered with brown punctures, sinuated and acuminated at the apex, and with three rows of distant large and shallow punctures on each elytron ; there are also on each elytron a lateral spot, and a narrow wavy subapical fascia, of an indistinct brown or black.

178.—HYDATICUS CONSANGUINEUS, Aubè. *Spec. gen. page* 160.

179.—HYDATICUS BIHAMATUS, Aubè. *Spec. page* 174. Eschsch. Dej. *Cat.* 3rd ed. page 61. *Goryi*, Aubè. *Spec. page* 175. *scriptus*, Blanch. *Voy. Pole. Sud.* iv., *p.* 46.

GYRINIDÆ.

180.—ENHYDRUS ORLONGUS, Boisd. *Voy. Astrol. Ent. page* 52. Aubè. *Spec. p.* 653. Dej. *Cat.* 3rd ed. p. 66. *Australis*, Brulle. *Hist. Nat. Ins.* v., *page* 237.

181.—GYRINUS OBLIQUATUS, Aubè. *Spec. page* 661.

182.— GYRINUS CONVEXIUSCULUS. n. sp.

Length 2 lines.

Black, nitid, convex, and rather elongate. Head with two shallow foveæ between the eyes, and the clypeus covered with fine striolæ. Thorax rounded behind with the anterior transversal line well marked and punctured near the sides, and a median transversal line well marked in the centre. Elytra with a broad depression close to the apex, which is rounded in each elytron, and with eleven light striæ marked with distinct punctures on each.

HYDROPHILIDÆ.

183.—HYDROPHILUS GAYNDAHENSIS. n. sp.

Length 12 lines.

This species may most readily be distinguished from the other Australians of the genus by the very great length of the sternal spine, which passes the extremities of the posterior thighs. The anterior tarsi, the hairs on all the tarsi, and the palpi are reddish.

184.—STERNOLOPHUS NITIDULUS. n. sp.

Length 5½ lines.

Black, very nitid, convex, and of an elongated oval form. Head with a shallow punctured impression on each side near the eyes, and with a semi-circular row of punctures extending from the anterior part of the eye forwards, and then inwards and backwards towards the centre of the forehead. Thorax broad, wider behind than in front, with all the angles rounded, and with two transverse punctured lines on each side, one near the apex, the other behind the middle. The elytra are rounded behind and very smooth, but there are faint traces of a few rows of punctures on them. The tarsi and palpi are red.

HYDATOTREPHIS. n. gen.

This genus, of which there are a number of species in Australia, has a general resemblance to *Hydrobius*. It differs however from that genus in having only 8 jointed antennæ, in not having the posterior tarsi ciliated, and in having the mesosternum prolonged into a tubercle in front of the intermediate thighs. The form of the body is short, oval, and convex. The maxillary palpi are short for the family, and the last joint is not longer than the penultimate. The rest as in *Hydrobius*.

185.—HYDATOTREPHIS MASTERSII. n. sp.

Length 4 lines.

Black, very nitid, very finely and closely punctured. Head with a broad shallow punctured depression on each side between and near the eyes. Thorax broad, slightly rounded on the sides, broader behind than in front, truncate at the base, with all the angles rounded, and with the apex and base narrowly, and the sides broadly, bordered with red. Elytra, with a deep stria on each side of the suture, commencing a little above the middle, and extending to the apex, and with several indistinct rows of irregular and distant punctures on the disk. The under sides of the thighs and sides of the metathorax are covered with a pale silky pubescence.

186.—PHILHYDRUS ELONGATULUS. n. sp.

Length 2 lines.

Elongate, ovate, nitid, subconvex, very finely punctured and of a lurid red clouded with brown. Head and anterior margin of thorax brown. Elytra with a distinct stria on each side of the suture not reaching to the base, and with obsolete brown punctured striæ over the rest of their surface.

187.—PHILHYDRUS MACULICEPS. n. sp.

Length 1¼ lines.

Short, ovate, convex, and subnitid. Head black with a large yellow spot on each side in front. Thorax luteous with the anterior border and a large central spot, brown. Elytra brown, bordered and clouded with yellow, and with a distinct stria on each side of the suture not reaching the base, and several series of obsolete punctured striæ.

188.—PHILHYDRUS MARMORATUS. n. sp.

Length ¾ of a line.

Short, ovate, convex, and subnitid. Head black with a purplish gloss. Thorax pale red with the middle brown. Elytra yellow clouded with brown, and with a number of narrow brown punctured striæ.

This may perhaps be a *Berosus*. I have not been able to examine the species satisfactorily.

HYDROBATICUS. n. gen.

But for the total absence of any tubercle or carina on the mesosternum, the two species I describe under this genus, might be almost as well placed with *Philhydrus*. They are also, however, of a more broadly ovate and flat form. As regards the antennæ and palpi they exactly correspond with *Philhydrus*.

189.— HYDROBATICUS TRISTIS. n. sp.

Length $2\frac{3}{4}$ lines.

Flat, opaque, coarsely punctured and of a dull reddish brown colour clouded with black. Head rather impressed along the suture of the epistome, with the labrum black. Thorax considerably broader than the length, slightly rounded on the sides and not broader behind than in front, with the anterior angles very round, the posterior angles obtuse, and the base truncate. Elytra broader than the thorax, broadly rounded at the apex, and covered with numerous coarse punctured striæ, and with the disk of a much darker colour than the sides.

190.—HYDROBATICUS LURIDUS. n. sp.

Length 2 lines.

Of the same form and sculpture as the last described species, differing only in size, and in being entirely of a pale lurid colour, with the exception of the labrum and back of the head which are black. In both this and the former species the scutellum is smaller and less elongate than is usual in *Philhydrus*.

I regret that I am unable from paucity of specimens to make a proper examination of either of the species.

HYGROTROPHUS. n. gen.

Head large, much prolonged in front of the eyes, and more or less inflexed beneath the thorax, and narrowed into a neck behind the eyes, which are prominent. Labrum rounded anteriorly, transverse. Mandibles large, strong, arcuated and bidentate on the inner side. Antennæ as in *Berosus*. Maxillary palpi not so long as the antennæ. Thorax rounded at the base. Scutellum elongate, triangular and acute at the apex. Elytra oval, convex. Anterior coxæ very large. Mesosternum very faintly carinated. Abdominal segments, six, the last very small.

This genus though in many respects like *Berosus*, differs from it in a very striking manner in the drooping head and rounded base of the thorax. The two species which I have put under it differ very much from one another in appearance.

191.—HYGROTROPHUS NUTANS. n. sp.

Length 2½ lines.

Of a lurid hue clouded with brown, closely punctured and covered with a very fine short pubescence. Thorax transverse, much rounded at the anterior angles, almost rectangular behind, with the base broadly and rather slightly rounded, and with the punctures on its disk transverse. Elytra with a number of rather lightly marked punctured striæ, and with the interstices flat. Posterior tarsi moderately ciliated. Under side of body black, legs yellow.

192.—HYGROTROPHUS INVOLUTUS. n. sp.

Length 1½ lines.

The very inflexed head, and convex arched form of this insect gives it a very different appearance from the last. The head and thorax are very closely and sharply punctured, and of a golden green colour, the latter narrowly bordered with yellow. Elytra very convex, pointed at the apex, and of a yellow colour, with ten well marked brown largely punctured striæ on each elytron. The whole under side and the basal half of the four posterior thighs are black and roughly punctured. The legs are yellow.

193.—HYDROCHUS PARALLELUS. n. sp.

Length 1¼ lines.

Of a dark bronzy green and opaque. Head punctured, with a deep stria on each side between the eyes extending from the occiput to the suture of the epistome, and with a shorter one in the middle not reaching the occiput. Thorax coarsely punctured, with the surface covered with several large shallow indistinct foveæ. Elytra more nitid than the thorax, with the suture broad and smooth, and on each side of it ten striæ, placed close together and filled with large punctures, and with the alternate interstices slightly elevated. The legs are reddish, the knees black.

194.—HYDRÆNA LURIDIPENNIS. n. sp.

Length ¾ of a line.

Elongate-ovate, subconvex, and closely punctured. Head black. Thorax pale at the sides, brown in the middle, not broader than the length, with all the angles acute, and with the sides bulged out almost angularly in the middle. Elytra of a lurid brown closely covered with punctate striæ.

195.—CYCLONOTUM MASTERSII. n. sp.

Length 2½ lines.

Oval, convex, black, subnitid, and entirely covered with fine punctures. The thorax is narrowly tinged with red on the lateral margins. The elytra have a narrow stria on each side of the suture. The under side is black and smooth. The tarsi and palpi are red.

196.—CYCLONOTUM PYGMÆUM. n. sp.

Length 1 line.

With the exception of the great difference in size, the greater proportionate width of the elytra behind, and the more oval oblong form, there is very little difference between this and the last described species. It has the same uniform black colour, the same kind of puncturation, and the same subsutural striæ on the elytra.

STAPHYLINIDÆ.

Sub-family ALEOCHARIDÆ.

MYRMECOCEPHALUS. n. gen.

Antennæ filiform, the first joint thick and of the same length as the third, the rest nearly equal and enlarging slightly towards the extremity. Maxillary palpi with the third joint much larger than the second, and the fourth very small and awl-shaped. Eyes of medium size, round and not prominent. Head convex, nearly square, subtruncate behind, rounded at the posterior angles, and affixed to the thorax by a short but very narrow neck. Thorax of an elongate oval form, narrower than the head, convex, and deeply impressed along the median line. Elytra broader than the head. Abdomen moderately elongate, widening a little towards the apex.

I have not been able to give a more complete detail than the above of the characters of this genus, for want of duplicate specimens, but enough I think has been given to enable it to be readily identified. The resemblance to an ant, especially about the head is very remarkable, and it is most probable that like some other genera of the Aleocharidæ its true habitat is in ants nests. The two species described below, were taken by Mr. Masters in a heap of debris left on the banks of the Burnett River after a heavy flood.

197.—MYRMECOCEPHALUS CINGULATUS. n. sp.

Length 1¼ lines.

Black, finely punctured, with the antennæ and palpi of a brownish yellow. The abdomen is subnitid with the hinder part of the first segment yellow.

198.—MYRMECOCEPHALUS BICINGULATUS. n. sp.

Length 1¾ lines.

This species differs from the last in having the head more rounded behind, in having the thorax more angularly rounded on the sides at its broadest part, in having the elytra of a more olive black, and in having the second as well as the first segment of the abdomen with a yellow ring.

199.—Tachyusa coracina. n. sp.

Length 1 line.

Black, nitid, and finely punctured, with the tarsi pale red. The elytra much wider than the thorax.

200.—Myrmedonia australis. n. sp.

Length 1¼ lines.

Of a pitchy red colour turning to brown on the elytra and extremity of the abdomen. The antennæ are very hairy, with the apical joints yellow. The whole body is nitid and very finely and closely punctured.

201.—Homalota flavicollis. n. sp.

Length ¾ of a line.

Head black. Thorax yellow. Elytra brown. Abdomen with the two first segments red, the third and fourth dark blue, and the remainder yellow.

202.—Homalota pallidipennis. n. sp.

Length ¾ of a line.

Entirely of a pale testaceous colour, with the exception of the third and fourth segments of the abdomen which are of a darker hue.

203.—Oxypoda analis. n. sp.

Length 2 lines.

Black, nitid, finely punctured, and clothed with a short ashen pubescence. The elytra have a tinge of pitchy red throughout, and the apex of the abdomen and the legs are red.

204.—Aleochara hæmorrhoidalis, Guer. Voy. Coquille. Ins. II, page 63, t. 1 f. 24.

205.—ALEOCHARA MASTERSII.   n. sp.

Length 3 lines.

This species differs from *A. hæmorrhoidalis*, in being much smaller, in the sculpture of the elytra, which instead of being smoothly punctured as in that species is rugose and covered with small striolæ, and in the absence of a red termination to the abdomen.

Sub-family TACHYPORIDÆ.

206.—CONURUS RUFIPALPIS.   n. sp.

Length 1½ lines.

Black, subnitid.   Thorax with minute striolæ at the base and sides.   Elytra finely punctured, marked with striolæ at the base which is of a deep red colour, and obliquely truncate at the apex. Abdomen short, and fimbriate.   Legs and palpi pale red.

207.—CONURUS ATRICEPS.   n. sp.

Length 1¼ lines.

Black, very nitid, smooth.   Thorax red.   Elytra of a darker red with the sides and base almost brown.   Abdomen short, fimbriate and with the terminal segment and the hinder margin of the others of a pitchy red.   The legs, antennæ, and palpi are red.

208.—CONURUS ELONGATULUS.   n. sp.

Length 2 lines.

Of an uniform dull black, and marked almost all over with minute striolæ.   The legs, antennæ, and palpi are of a brownish yellow.   This species differs considerably in form from the two described above, it is less convex, proportionately narrower, and has the abdomen much more elongate.

209.—TACHYPORUS TRISTIS.   n. sp.

Length 1 line.

Head, thorax, and elytra dark brown, the two former very finely punctured, the latter punctured on the anterior half, striolate on the posterior, and slightly conjointly emarginate at the apex. The abdomen is black, the segments finely striolated. The legs and antennæ are brownish yellow.

210.—TACHYPORUS RUBRICOLLIS. n. sp.

Length 1 line.

Head and thorax red, subnitid. Elytra pitchy brown and covered with fine striolæ. Body black, legs and antennæ red.

Sub-family STAPHYLINIDÆ.

211.—LEPTACINUS LURIDIPENNIS. n. sp.

Length 2 lines.

Head black, nitid, parallel-sided, much longer than the breadth, a little rounded at the posterior angles, and marked with distant punctures. Thorax dark red, nitid, longer and a little narrower than the head, slightly emarginate on the sides behind the middle, very slightly narrower behind than in front, rounded at both ends and sparingly marked with punctures. Scutellum large, triangular, and of a brown colour. Elytra roughly punctured, broader than the thorax, rounded at the apex, of a brown colour at the base, and of a pale nitid lurid hue on the posterior half. Abdomen dark brown and punctate. Legs, antennæ, and palpi brownish red.

The whole insect has an elongated flat appearance.

212.—LEPTACINUS CYANEIPENNIS. n. sp.

Length 3 lines.

Head black, nitid, parallel-sided, twice as long as the width, truncate behind, with a deep margined impression on each side extending from the inner side of the eye in an oblique direction towards the centre of the forehead, and with a few large punctures on the back part of the head and towards the labrum. Neck very slight. Thorax bright red, nitid, shaped much as in the last described species, but more convex, and with the punctures more rare. Scutellum red. Elytra dark blue and finely punctured.   Abdomen black, nitid, smooth and rather convex, with the first and fourth segments red.   The legs also are red.

213.—XANTHOLINUS ATRICEPS.   n. sp.

Length 2¾ lines.

Head black, nitid, smooth, a little longer than the breadth and rounded behind, with a few punctures scattered over the surface.   Thorax red, sparingly punctured, longer than the breadth, broader than the head, slightly broader behind than in front, rounded at both ends and very slightly rounded on the sides.   Elytra broader than the thorax, of a nitid brownish red colour, sparingly punctured and conjointly emarginated to a slight degree at the apex.   Abdomen brown, very minutely punctured, clothed with a very short fine pubescence, and having the last segment long, pointed, and of a dull red colour.   The legs are pale red.

214.—XANTHOLINUS PICEUS.   n. sp.

Length 2½ lines.

This species only differs from the last in being a little smaller and of an almost uniform pitchy brown hue.

215.—XANTHOLINUS CERVINIPENNIS.   n. sp.

Length 6 lines.

Black, subnitid.   Head nearly square, rather convex, truncate behind, and marked with a shallow foveæ on each side near the eyes, with nearer the middle two slightly oblique lines converging behind, and with a few punctures on the occiput and sides.   Thorax narrower than the head, longer than the breadth, truncate in front, narrowed a little behind, and rounded at the apex, with a few large punctures on the upper surface.   Elytra not broader than the broadest part of the thorax, coarsely punctured and of a pale brownish red colour.   The abdomen is thinly punctured, is clothed with soft hairs, and has the terminal segment and the hinder half of the penultimate one of a deep red.   The legs and palpi are piceous.

216.—XANTHOLINUS CYANEIPENNIS. n. sp.

Length 6 lines.

This species is about the same length as the last, but has a slighter and more elongate look. The head is not so broad, not so truncate behind, is proportionately longer, has the punctures between the eyes more numerous, and the oblique lines much shorter. The thorax is as broad as the head at the apex, which is truncate, but is somewhat narrowed towards the base which is rounded. The elytra are scarcely wider than the thorax, and are of a chalybeate blue, with a series of small punctures on each side of the suture, and another series in an obsolete depression in the middle of each elytron. The abdomen is dark olive, finely punctured and clothed with long soft hair. The legs and palpi are of a piceous red.

217.—XANTHOLINUS DUBIUS. n. sp.

Length 2¼ lines.

Of a pitchy black, nitid. Head longer than the breadth, and rather rounded behind. Thorax rounded in front and behind, and broadest at the base. Elytra sparingly punctured, broader than the thorax, and with a reddish tinge throughout. Abdomen rather dull, with a tinge of red on the two terminal segments. Legs red.

The last joint of the maxillary palpi in this species is neither so acuminated nor so slight as is usual in this genus. The form of the thorax also, which is very distinctly rounded at the anterior angles, marks it out as being a very aberrant species.

218.—PHILONTHUS AUSTRALIS. n. sp.

Length 5 lines.

Black, subnitid. Head almost round, with a few large setigerous punctures scattered over its surface. Thorax as broad as the head at the apex which is truncate, narrowed a little and rounded behind, longer than the breadth, and marked with a number of large punctures on each side. Scutellum large, triangular and closely punctured. Elytra broader than the thorax.

coarsely punctured, conjointly emarginate at the apex, and of a deep red colour. Abdomen punctate, and clothed with strong hairs, with the terminal segment and part of the penultimate red, and with the other segments tinged with blue. The tibiæ, tarsi, palpi, and three terminal joints of the antennæ are yellow.

219.—PHILONTHUS HÆMORRHOIDALIS. n. sp.

Length 5½ lines.

Head of a dark olive hue, very nitid, almost circular, and depressed, with a short longitudinal impression in the middle and two large punctures on each side of the forehead. Thorax marked and shaped as in the preceding species, but with the anterior angles quite square. Scutellum large, punctate and of a triangular form, with the sides a little rounded. Elytra punctate, broader than the thorax, rounded behind and conjointly emarginate, and of a brassy olive hue. Abdomen black, and finely punctured with the terminal segment and hinder portion of the penultimate one red. The legs and palpi are red. The antennæ have the three basal joints red, the three apical ones yellow, and the remainder black.

220.—PHILONTHUS PILIPENNIS. n. sp.

Length 3 lines.

Black, subnitid. Head nearly square, with a few punctures on each side. Thorax almost as broad as the head, longer than the breadth, almost parallel-sided, with the anterior angles square, the base round, and the surface smooth with the exception of a line of about six punctures on each side of the middle and a few scattered ones near the sides. Elytra broader than the thorax, punctate, and covered thinly with a greyish decumbent pile. Abdomen closely punctured and also clothed with a decumbent greyish pile. Legs, palpi, and basal joint of antennæ ferruginous.

221.—PHILONTHUS POLITULUS. n. sp.

Length 2½ lines.

Black, nitid. Head nearly circular and smooth. Thorax as broad as the head, longer than the breadth, and almost parallel-sided. Elytra broader than the thorax, thinly punctured, and with decumbent hairs. Abdomen also thinly punctured, with the anus moderately fimbriate. The legs and terminal joint of the antennæ are reddish.

222.—PHILONTHUS SUBCINGULATUS. n. sp.

Length 2½ lines.

Head black, nitid, smooth. Thorax red, nitid, rather longer than the breadth, broader behind than in front, rounded at the base and marked with a few punctures on the upper surface. Elytra broader than the thorax, punctured, and of a brown colour with the apical edge yellowish red. Abdomen black, opaque, with the apical edge of the three first segments tinged slightly with red. The legs and palpi are dark red, the antennæ brown.

223.—PHILONTHUS CHALYBEIPENNIS. n. sp.

Length 2½ lines.

Head and thorax nitid, the former black and nearly circular, the latter red, nearly square, slightly rounded on the sides, not broader behind than in front, and quite smooth. Elytra broader than the thorax, blue, punctured, and not nitid. Abdomen black, subnitid, and marked with small elongate punctures. The tarsi and three terminal joints of the antennæ are of a pale red.

224.—PHILONTHUS XANTHOLINOIDES. n. sp.

Length 1¼ lines.

This insect is entirely of a brownish red, with the exception of the head which is black. The thorax is nitid, considerably longer than the breadth, not broader than the head, broader behind than in front, and marked with two or three punctures on the upper surface. The elytra and abdomen are of a darker hue than the thorax. The two basal joints of the antennæ are red, the rest brown. The fourth joint of the maxillary palpi is long and unusually slight for the genus.

225.—STAPHYLINUS LURIDIPENNIS.   n. sp.

Length 3 lines.

Head and thorax black, nitid, the former somewhat small and
with a few punctures near the eyes and on the occiput, the latter
broader than the head, transverse, much widened and rounded
behind, and marked with two punctures near the middle of the
upper surface, and a few others along the basal margin.   Elytra
of a lurid hue, punctate, clothed with decumbent pile, and conjointly emarginated at the base.   Abdomen brownish black,
punctate and setose, with the apex reddish.   The tibiæ, tarsi,
palpi, and antennæ are also red.

226.—STAPHYLINUS ANALIS.   n. sp.

Length 3 lines.

This species only differs from the last, in having the elytra
quite black, and in having the apical portion of the abdominal
segments of a steel blue tint.

227.—CREOPHILUS ERYTHROCEPHALUS, Fab.   Syst. Ent.
page 265.   Oliv. Ent. III. 42. p. 12. t. 2.
fig. 9.   Erichs. Gen. page 351.

Sub-family PÆDERIDÆ.

228.—CRYPTOBIUM MASTERSII.   n. sp.

Length 5½ lines.

Head oblong, rounded at the posterior angles, black, opaque,
and closely punctured.   Thorax red, subnitid, sparsely punctured,
narrower than the head, much longer than the breadth, parallel-
sided and with all the angles rounded.   Elytra broader than the
thorax, conjointly emarginate behind, densely punctured, black
on the basal and larger half, and deep red on the apical portion.
Abdomen black and finely punctured.   Thighs yellow with
black tips.   Tibiæ, tarsi, and palpi dull red.

229.—CRYPTORIUM APICALE. n. sp.

Length 3½ lines.

Head oblong, black, opaque, closely punctured, and truncate behind, with the posterior angles rounded. Thorax black, opaque, closely punctured, narrower than the head, much longer than the breadth, and slightly rounded on the sides, with all the angles rounded, and the median line smooth. Elytra broader than the thorax, conjointly emarginate at the apex, closely punctured, and of a black colour with the hinder portion deep red. The abdomen is black, subnitid, and finely punctured, with the fifth segment of a blood red. The legs are yellow. The antennæ, palpi, and mandibles are reddish brown.

230.—DOLICAON QUADRATICOLLIS. n. sp.

Length 3 lines.

Head black, nitid, square, truncate at the base, and covered with rather distant punctures. Thorax red, nearly square, and rounded at the angles, with the surface sparsely punctured, and the median line smooth. Elytra a little broader than the thorax, black with the apex red, and punctured in somewhat irregular rows over the whole upper surface. Abdomen black and very finely punctured, with the apex red and fimbriate. Legs, palpi and antennæ pale red.

231.—DOLICAON ELONGATULUS. n. sp.

Length 2½ lines.

Black, subnitid. Head more elongate than in the last species and more closely punctured. Thorax considerably longer than the breadth, slightly narrower than the head, rather closely punctured on the upper surface, rounded at the angles, and very slightly broader in front than behind. Elytra broader than the thorax, punctured in somewhat irregular rows with subelevated interstices, and with the base black and apex red. Abdomen black and finely punctured, with the apex red and fimbriate The legs, antennæ, and palpi, reddish yellow.

232.—DOLICAON NIGRIPENNIS.   n. sp.

Length 3 lines.

Head black, nitid, and very sparsely punctured. Thorax red, nitid, same width as head, longer than the breadth, parallel-sided, rounded at the angles, and very sparsely punctured. Elytra broader than the thorax, black, and punctate. Abdomen very finely punctured, and broadest in the middle, with the first four segments red, and the remainder black. Legs, palpi, antennæ, and mandibles red.

233.—LATHROBIUM POLITULUM.   n. sp.

Length 5 lines.

Black, hairy, and nitid. Head slightly narrowed and rounded behind, punctured sparsely in front and more closely on the sides. Thorax of the same width as the head, a little longer than the breadth, slightly rounded on the sides, rounded at the angles, and sparsely punctured on the upper surface. Elytra broader than the thorax and punctured in rows. Abdomen finely punctured with the apical margin of each segment smoother and of a lighter colour. Legs, antennæ, and palpi, red.

234.—LATHROBIUM PICEUM.   n. sp.

Length 6½ lines.

Entirely of a piceous brown, subnitid, hairy. Head and thorax of the same form as in L. *politulum*, but with the punctures smaller and more numerous. Elytra not broader than the thorax, and closely punctured. Abdomen rugosely punctured with a reddish tinge on the apical margin of the first, third, and fourth segments. The antennæ and palpi are nearly red, the latter are considerably thicker than in the previously described species.

235.—LITHOCHARIS TRISTIS.   n. sp.

Length 1¾ lines.

Black, opaque, and punctate. Head square, truncate behind and placed on a narrow neck. Thorax reddish brown, longer than the breadth, not broader than the head, and with all the angles rounded. Elytra dark brown, and broader than the thorax. Abdomen finely punctured and striolate. Legs, antennæ, and palpi, reddish brown.

236.—SCOPÆUS ROTUNDICOLLIS.   n. sp.

Length 1½ lines.

Pale brown, strongly and closely punctured and thinly clothed with long hairs. Head large, flat, truncate behind, and with the posterior angles rounded. Thorax a little longer than the breadth, narrower than the head, much rounded on the sides, and rounded and narrowed at the apex and base. Elytra broader than the thorax, with the apical half of a reddish yellow. Abdomen black, broad, rather short, with the fourth segment reddish. The legs, antennæ, and palpi, yellow.

237.—STILICUS OVICOLLIS.   n. sp.

Length 1¼ lines.

Pale red, nitid, smooth. Head square, truncate behind, and joined to the thorax by a very slight neck. Thorax of an elongate oval form, with a peduncular attachment to the body. Elytra very little broader than the thorax, with the basal portions brown. Abdomen with the first four segments brown.

238.—SUNIUS CYLINDRICUS.   n. sp.

Length 2 lines.

Head black, densely punctured, square, and truncate behind. Thorax red, closely punctured, narrower than the head, and of a short oval form. Elytra scarcely broader than the thorax, punctate, and of a brownish black colour with the base and apex pale red. Abdomen elongate, subcylindrical, a little widened towards the apex, of a reddish colour, with the fifth and part of the sixth segments black.

This species seems to vary a good deal in colouring. The thorax and abdomen have brown marks in one of the specimens before me which are not to be traced in the other.

230.—PÆDERUS CINGULATUS. n. sp.

Length 3½ lines.

This is evidently a distinct species from *P. Australis* Guer., and *P. cruenticollis* Germ., the only Australian species described. I cannot, however, find any positive difference excepting in the coloration of the abdomen. In the present species the third and fourth segments are entirely red.

240.—PÆDERUS ANGULICOLLIS. n. sp.

Length 2¼ lines.

This insect is much smaller than the last. It may be readily distinguished by the almost square thorax and by the four basal segments of the abdomen being of a dull red.

Sub-family PINOPHILIDÆ.

241.—PINOPHILUS GRANDICEPS. n. sp.

Length 7 lines.

Black, subnitid, closely punctured, and clothed with a soft ashen pile. Head sparsely punctured in front, broad, slightly convex, truncate behind, and affixed to the thorax by a thick but very distinct neck. Thorax of the same width as the head in front, scarcely longer than the breadth, and slightly narrowed towards the posterior angles which are rounded. Elytra not so broad as the thorax at its broadest part and very short. Abdomen elongate, somewhat flat, and becoming pointed at the apex. The tarsi and antennæ are reddish, the latter are of a very slight form. The last joint of the maxillary palpi is large, pointed, and directed inwards, but it scarcely answers to the description of the genus given by Lacordaire.

242.—PINOPHILUS MASTERSII. n. sp.

Length 6½ lines.

This species may be readily distinguished from the last by the much finer and denser puncturation over the whole body, and the denser clothing of soft ashen pile. The thorax also is slightly longer, the extremity of the abdomen is of a reddish tint, and the legs are entirely of a light red. The fourth joint of the tarsi is not so strongly bilobed as in *P. grandiceps*.

243.—PINOPHILUS BREVIS. n. sp.

Length 3 lines.

Though differing very much in size, this insect much resembles *P. grandiceps*. The thorax however is more square, and is very briefly carinated at the base on the median line. The legs are entirely red.

244.—OEDICHIRUS PÆDEROIDES. n. sp.

Length 3½ lines.

Black, nitid. Thorax red, longer than the breadth, rounded behind the anterior angles, narrowed at the base, and marked with two deeply punctured impressions not reaching the apex and some lateral punctures. Elytra broader than the thorax, emarginate at the apex, and punctured, with the punctures large and scattered. Abdomen elongate, cylindrical, and punctured in transverse rows four on each segment, with the three first segments red, the rest black. The maxillary palpi are black, with the last joint very much prolonged internally.

PINOBIUS. n. gen.

Antennæ short, moniliform, somewhat geniculate, first joint thick, the last pointed. Maxillary palpi with the last joint large, subsecuriform, laterally compressed, and very slight at its junction with the penultimate one, which is as long, but much slighter and a little arcuated. Head round, and depressed. Neck broad, distinct. Thorax nearly square. Elytra narrow. Abdomen large, elongate, and cylindrical, with the lateral border feebly marked. Legs rather slight, anterior tarsi moderately dilated with the fourth joint of all small, and slightly lobed.

Unfortunately there is only one specimen and that a bad one, of the insect for whose reception this genus is formed. The description is consequently very imperfect, but until specimens can be procured it must suffice. I am far from sure that I am right in putting it with the *Pinophilidæ*, but it is marvellously unlike any genus of the *Pæderidæ*, the only other division of the Brachylytra, which it can be associated with.

245.—PINOBIUS MASTERSII. n. sp.

Length 6½ lines.

Head thinly punctured and subnitid, with the front part black and the rest red. Neck red. Thorax not broader than the head, a little longer than the breadth, not narrowed behind, thinly punctured, and red with black base. Elytra thinly punctured, and black, with a large blood red spot in the centre of each elytron. Abdomen broader than the elytra, and finely punctured, with the two apical segments black, the rest red. The tarsi and two terminal joints of the antennæ are reddish.

Sub-family STENIDÆ.

246.—STENUS MACULATUS. n. sp.

Length 3 lines.

Black, roughly punctured. The thorax has the median line deeply marked. On each elytron there is a large golden yellow spot, nearer to the side than to the suture, and nearer to the base than to the apex. The legs are red with the knees black.

247.—STENUS GAYNDAHENSIS. n. sp.

Length 2 lines.

Black, strongly punctured. Head with a short raised smooth line on the middle and a longer one on each side between the eyes. Thorax nearly cylindrical, with a smooth raised central line. Legs yellow, with the knees brown.

248.—STENUS OLIVACEUS. n. sp.

Length 2 lines.

This species differs from the last in having the head much

excavated between the eyes, and without any smooth ridges, in having the thorax more bulged out on the sides, and with an olive gloss both on it and the elytra. The legs are black, with the exception of the upper part of the thighs which are yellow.

249.—STENUS SIMILIS. n. sp.

Length 1½ lines.

Only differs from the last in being much smaller, in having the head less excavated between the eyes, in having the thorax more thinly punctured, and in wanting the olive gloss on both thorax and elytra.

250.—STENUS VIRIDIÆNEUS. n. sp.

Length 2 lines.

Black, opaque. Head smooth between the eyes. Thorax without puncturation in the middle, but with transverse looking punctures towards the apex and base, which are both rather constricted. Elytra strongly punctured, and of a brassy green colour with a ruddy hue in the middle.

251.—STENUS CUPREIPENNIS. n. sp.

Length 2½ lines.

Head and thorax lightly punctured, the former with a greenish the latter with a bluish gloss. Elytra punctured and of a brilliant coppery red. Abdomen bluish black, nitid. Basal half of the thighs yellow, the rest brown. Fourth joint of the palpi not bilobed.

252.—STENUS PUNCTICOLLIS. n. sp.

Length 2 lines.

This species also has the elytra of a coppery red, but of a redder and less brilliant character than in the last described insect. The head and thorax are closely and evenly punctured all over, in this respect differing much from both the preceding species. In other respects it exactly resembles *S. viridiæneus.*

Sub-family OXYTELIDÆ.

253.—MEGALOPS NODIPENNIS. n. sp.

Length 1½ lines.

Pale red, subnitid. Head large and vertical, with a circular impression between the eyes, and large punctures over the whole surface. Thorax nearly as broad as the head, not longer than the breadth, subcylindrical, and covered with a number of deep transverse impressions full of punctures, giving a rough and irregular appearance to the whole. Elytra brown near the suture, the remainder occupied by large yellow rounded nodular elevations. Abdomen thick, short, and rounded at the apex.

254.—BLEDIUS MANDIBULARIS. n. sp.

Length 1½ lines.

Brown, opaque, finely punctured. Head with a small horn on each side near the eyes. Thorax nearly square, and rounded at the posterior angles, with a very fine median line. Elytra a little broader than the thorax, conjointly emarginate and slightly dehiscent. Abdomen black, with the tip yellow. Legs and palpi reddish.

255.—OXYTELUS BRUNNEIPENNIS. n. sp.

Length 1¾ lines.

Red clouded with brown, subnitid. Head large, smooth, excavated in front, terminating in a sharp point before the eyes, truncate behind, and attached to the thorax by a thick neck. Thorax transverse, truncate in front, rather rounded behind, with the median line well marked, and on each side two deep longitudinal grooves, the external one near the lateral margin. Elytra broad, truncate behind. Abdomen broad, rather depressed, and becoming suddenly pointed at the apex.

256.—OXYTELUS IMPRESSIFRONS. n. sp.

Length 1½ lines.

Black, nitid. Head large, smooth, and truncate, with a short deep impression in the centre of the forehead. Thorax the same as in the last species, but with the grooves less deeply marked. Elytra brownish yellow. Abdomen as in the last species. Legs yellow.

Sub-family PIESTIDÆ.

257.—ISOMALUS PLANICOLLIS. n. sp.

Length 2 lines.

Black, nitid, very flat and smooth. Head large, square and fitting to the thorax. Thorax triangular, the apex towards the abdomen. Elytra longer than the breadth, of a piceous hue and with a puncture in the middle of each elytron. Abdomen with the apical edge of the second and third segments slightly red. The legs, antennæ, palpi, and mandibles are piceous.

Sub-family OMALIDÆ.

258.—OMALIUM GAYNDAHENSE. n. sp.

Length 1 line.

Black, subnitid. Head and thorax very finely punctured, the latter transverse, truncate in front and behind, and slightly rounded on the sides. Elytra more closely punctured, and of a dark brown colour. Abdomen broad and rounded at the apex. Legs red. Antennæ brown.

PSELAPHIDÆ.

259.—TMESIPHORUS KINGII. n. sp.

Length 1¼ lines.

Brown, opaque, and roughly punctured. Antennæ with the ninth and tenth joints large, and oblong, and with the eleventh still larger and round. Head with a deep groove in front between the antennæ, and two large punctures on the forehead. Thorax gibbous. Elytra red and finely punctured, with a deep longitudinal impression at the base on each side midway between the suture and humeral angle. Abdomen finely punctured, bordered laterally and bicarinated, the carina not extending beyond the second segment.

This species, which would appear from the numbers captured to be abundant about Gayndah, very much resembles *T. Macleaii* King, it is, however, of a very much darker colour, and is altogether a rougher and more strongly sculptured insect.

260.—TYRUS MASTERSII.   n. sp.

Length 1 line.

Red, subnitid and punctured. The second joint of the maxillary palpi is long, slight at the base and much enlarged at the apex, with a sharp protuberance on the inner side at the thickest part; the third is short and turbinate, with an angular protuberance on the inner side, and the fourth is large, ovate and somewhat acuminated. The ninth and tenth joints of the antennæ are a little longer than the preceding ones, and the eleventh is large and oval. The head is rather prolonged in front. The thorax is scarcely longer than the breadth, very little narrowed behind and truncate. The elytra and abdomen are covered with a fine ashen pubescence ; the former are of a lighter red than the rest of the body, and have the basal striæ short and light.

The maxillary palpi of this species are unlike those of any *Tyrus* I have seen, but whether the divergence from the typical form is sufficient to render the formation of another genus necessary, is a question I must leave to be decided by those who have made this Family their peculiar study.

261.—BRYAXIS HIRTA.   n. sp.

Length 1 line.

Red, subnitid, punctate, and clothed with ashen pubescence. Antennæ with the terminal joint large and pointed, and the 9th and 10th joints a little longer than the preceding ones. Head with two deep foveæ on the forehead. Thorax much widened at the sides so as to form an angle near the middle, and narrowed behind, with the posterior angles acute, and the median line lightly carinated. Elytra convex and without striæ.

262.—BRYAXIS ATRICEPS.   n. sp.

Length 1 line.

Red, nitid, and smooth. Antennæ long, with the ninth joint

a little larger than the preceding ones, the tenth elongate, and the eleventh of a very elongate oval form. Head black and flat on the forehead with two deep round foveæ. Thorax convex, and rounded at the sides, with a deep fovea on each side near the base connected together by a transverse impression. Elytra bistriated, one on each elytron. Legs long.

PAUSSIDÆ.

263.—ARTHROPTERUS WESTWOODII. n. sp.

Length 5½ lines.

Piceous brown, subnitid, and finely punctured. Antennæ short, with the first joint transverse, obtusely angled, and truncate, the second to the ninth inclusive more than four times broader than the length, and the tenth more than twice the length of the preceding. Head slightly concave between the eyes, truncate behind, and attached to the thorax by a thick neck, with the posterior angles prominent, obtuse, and clothed with stiff hairs. Thorax scarcely longer than the breadth, rounded at the anterior angles, slightly narrowed behind, truncate at the base, margined and ciliated on the sides and broadly impressed on the median line, more especially towards the base. Elytra rounded at the apex with a small sinuation at the external angle. Legs and under side of body thinly punctured. Anterior tibiæ with the external apical angle subacute and the apex deeply emarginate. Intermediate and posterior tibiæ with the external apical angle very broadly rounded and ciliated, and with the apex emarginate.

264.—ARTHROPTERUS MASTERSII. n. sp.

Length 6½ lines.

Piceous-black, subnitid, and finely punctured. Antennæ with the first joint square, obtusely angled, and truncate, the second to the ninth inclusive three times broader than the length, and the last more than twice the length of the others. Head slightly concave between the eyes, truncate behind, and attached to the thorax by a thick neck, with the posterior angles obtuse. Thorax much longer than the breadth, and very slightly narrowed behind, with the median line deeply impressed in the middle, but not extending to the apical and basal margins, and with an indistinct fovea near the basal part of the lateral margin. Elytra truncate at the apex and slightly notched at the external angles. Legs and underside of body closely punctured. The four anterior tibiæ have the external apical angle acute, and the apex deeply emarginate, the posterior are rather more obtusely angled.

265.—ARTHROPTERUS ANGUSTICORNIS.   n. sp.

Length 5½ lines.

Of a piceous brown colour and nitid. Antennæ narrow, the first joint square with the angles obtuse, the second to the ninth inclusive twice as broad as the length, the last equal in length to the two preceding united. Head depressed on the vertex and coarsely punctured, with a prominent ciliated tubercle at the posterior angles, and the base truncate. Thorax subcordiform, coarsely and transversely punctured, and with the median line lightly impressed in the middle. Elytra thinly and finely punctured, and truncate at the apex, with two small notches at each external angle. Body beneath and legs thinly punctured. All the tibiæ have the external apical angle very acute.

This species seems to approach the *A. parallelocerus* of Westwood.

266.—ARTHROPTERUS KINGII.   n. sp.

Length 5½ lines.

This species differs from the last in being of a pitchy red colour, in having no depression on the top of the head, in the tubercle at the posterior angles of the head being much smaller, in the thorax being more ciliated on the sides and having the median line more marked, in the external apical angles of the tibiæ being less acute, and in the narrower form of the whole body. The antennæ are of the same character as those of *A. angusticornis*.

267.—ARTHROPTERUS ELONGATULUS.   n. sp.

Length 4 lines.

Long, narrow, of a reddish colour, subnitid and punctate.

Antennæ rather short, the first joint nearly square, and with the angles obtuse, the second to the ninth inclusive three times broader than the length, and the last more than twice the length of the preceding one. Head not depressed in front, and with the posterior angles rounded. Thorax much longer than the width, very little narrowed behind, and not rounded on the sides, with the median line obsoletely carinated and broadly depressed near the base. Elytra a little wider than the thorax, and with the apex triemarginate in nearly equal lengths, and produced into acute points between the emarginations. Apex of the abdomen nearly black. The external apical angle of the fore tibiæ is acute, that of the intermediate and posterior tibiæ is rather obtuse, with the apex subtruncate.

SCYDMÆNIDÆ.

268.—Phagonophana Kingii, King. *Trans. Ent. Soc. N. S. Wales, Vol.* 1, *page* 92.

269.—Scydmænus Kingii. n. sp.

Length ¾ of a line.

Dark red, nitid and clothed with a short golden pubescence. Antennæ of medium size, with the two last joints forming an elongated club, and the ninth joint a little larger than the preceding ones. Neck distinct. Thorax rather elongate, not constricted behind. Thighs long, clavate.

SILPHIDÆ.

270.—Ptomaphila lachymosa, Schreib. *Trans. Linn. Soc., Vol.* VI., *page* 194.

271.—Catops obscurus. n. sp.

Length 1¼ lines.

This species differs from *C. australis* Erichs., the only Australian species hitherto described, in being clothed with a light coloured pubescence, and in having the thorax and elytra longitudinally instead of transversely scratched or striolated.

SCAPHIDIIDÆ.

272.—SCAPHIDIUM PUNCTIPENNE. n. sp.

Length 1¾ lines.

Black, nitid. Thorax broadly margined with red excepting in the middle of the base, and with a transverse row of deep punctures near the basal margin. Elytra, with a row of deep punctures near the base from the suture to near the humeral angles, with two or three longitudinal rows of large and rather distant punctures, and with two red spots on each elytron, one, transverse and near the base, the other, smaller and near the apex. The segments of the abdomen showing beyond the elytra are red with the apex black. The middle of the posterior thighs is red.

273.—SCAPHIDIUM MASTERSII. n. sp.

Length 1½ lines.

This species is smaller than the last, and differs from it in having only the sides of the thorax red, in having only one red spot on each elytron situated near the centre and of a transverse shape, in having the rows of punctures on the elytra more large and distinct, and in having all the abdominal segments red, with a black spot on the pical one.

274.—SCAPHISOMA POLITUM. n. s.p.

Length ¾ of a line.

Black, nitid and smooth, with a light stria on each side of the suture of the elytra, with the abdomen of a pitchy red, and with the legs and antennæ of a pale red.

275.—SCAPHISOMA PUNCTIPENNE. n. sp.

Length ⅓ of a line.

Black and nitid, with the elytra and meso and meta-thorax rather thickly punctured, with a light stria on each side of the suture of the elytra, and with the legs red.

HISTERIDÆ.

276.—HOLOLEPTA MASTERSII. n. sp.

Length 5 lines.

This species is very like *H. Sidnensis* Mars. It is of a slightly thicker and less flat form, is less punctured on the sides of the thorax, has the foveæ at the anterior angles of the thorax in the male shorter and rounder, and has the pygidium more punctate.

Mr. Masters found this insect in tolerable abundance in dead Bottle trees *(Sterculia ruprestris* Bentham), and only in them. The other two Australian species *H. Australis* Mars. from Western Australia, and *H. Sidnensis* Mars. from the neighbourhood of Sydney, are found exclusively in various species of grass tree *(Xanthorrhœa.)*

277.—PLATYSOMA SUBDEPRESSUM. n. sp.

Length $2\frac{1}{4}$ lines.

Black, nitid, and somewhat flat. Head excavated and punctured in front. Thorax finely punctured towards the sides. Elytra, with the stria nearest the suture extending from the apex to a third of the length of the elytra, with the second extending to one half the length, with the next three entire, and with a short oblique one near the humeral angle. Pygidium covered with variolous punctures.

278.—PLATYSOMA CONVEXIUSCULUM. n. sp.

Length $1\frac{1}{2}$ lines.

Black, nitid, and rather convex. Head excavated in front, and smooth. Thorax with a sublateral stria, and a small round depression in the centre of the base. Elytra with the first two striæ from the suture of equal length, extending from the apex half way up the elytra, and with the next three entire. Pygidium punctured as in the last species.

279.—PLATYSOMA PLANICEPS. n. sp.

Length 1 line.

Black, nitid, and convex. Head flat in front, with the suture

of the epistome straight and well marked. Thorax with a
faint sublateral stria. Elytra striated as in the last species.
Pygidium very thinly punctured.

The elongation and expansion of the prosternum under the
mouth is not nearly so great in this as in the two preceding
species, or as it is in the genus *Platysoma* generally. In the
absence, however, of specimens for minute examination and
dissection, I cannot venture to form a genus for its reception.

280.—SAPRINUS GAYNDAHENSIS. n. sp.

Length 2½ lines.

Head black, very minutely punctured, and with a distinct
uninterrupted stria surrounding the whole space between the eyes.
Thorax of a bronzy olive hue, nitid in the middle, and roughly
punctured towards the sides and along the basal margin, with an
impressed point in the centre of the basal lobe. Elytra blue,
nitid, and punctured behind and at the sides, with the sutural
stria not reaching the base, the next stria short, curved, punctured,
and taking its rise near the scutellum, the next short and parallel
to the rest, the fourth and fifth about half the length of the elytra,
and the sixth a little longer and almost joining the sublateral stria
which extends from the humeral angle to the apex where it
merges in the marginal one, and with a short stria on the side of
the humeral angle having the same basal origin as the sublateral
stria. Body beneath black and nitid. Legs piceous black.

281.—SAPRINUS MASTERSII. n. sp.

Length 2 lines.

Head duller and more punctured than in the preceding species.
Thorax black, with a slight coppery red reflexion, variolous punctures near the sides and along the base, and two shallow depressions near the apex at the commencement of the puncturated
patches. Elytra greenish black, with the apical two thirds punctured, the sutural stria not reaching the base, the first dorsal stria
about one third of the length of the elytra, the second third and
fourth half the length, the fifth long and almost joining the sublateral stria, the abbreviated humeral and sublateral stria as in *S.*

Gaymilahensis, and the interstice between the third and fourth striæ rugose and striolate.   Under side of body greenish black and punctured.

282.—ABRÆUS AUSTRALIS.   n. sp.

Length 1¼ lines.

Black and subnitid.   Elytra punctured, with a fine stria on each side of the suture, extending from the apex to near the base, and diverging gradually from the suture as it proceeds upwards.

NITIDULIDÆ.

283.—BRACHYPEPLUS MURRAYI.   n. sp.

Length 1¾ lines.

Like *B. binotatus* Murray.   Narrow, black, nitid.   Antennæ with the club pale rufous.   Head punctate and bifoveolate. Thorax slightly transverse, with the sides ciliated and reddish towards the posterior angles.   Elytra longer than the thorax, striate, finely punctured and pubescent, with a broad band of red at the base not extending to the suture.   The exposed segments of the abdomen are red with the sides black.

284.—CARPOPHILUS CONVEXIUSCULUS.   n. sp.

Length 1 line.

Red, subconvex, and finely punctate.   Antennæ with the terminal joint of the club yellow, the others brown.   Head much narrowed behind the eyes, which are prominent.   Thorax short, transverse, emarginate in front, truncate behind, broader at the base than at the apex, rounded at the anterior, and square at the posterior angles, with a broad black vitta in the centre, which is broadest at the apex, and is gradually narrowed to the base. Scutellum black.   Elytra of the same width as the thorax at the base, separately rounded at the apex, and of a subnitid black colour, with a basal fascia enclosing a black spot near each shoulder, and a round spot behind the middle, red.   Abdomen pointed at the apex.

K

This species ought perhaps to be placed in the genus *Stauro-glossicus* Murray. I have not been able to get a specimen for dissection and cannot therefore speak with certainty.

285. — CARPOPHILUS LURIDIPENNIS. n. sp.

Length ¾ of a line.

Head dark brown, punctate, and narrowed behind the eyes, which are prominent. Antennæ red with the club black. Thorax dull red, transverse, not broader behind than in front, punctate, and obtusely rectangular. Scutellum large, black, and triangular. Elytra not much longer than the breadth, not narrowed behind, truncate at the apex, punctate and of a dark lurid colour with the apex black. Abdomen reddish brown. Legs red.

286.—CARPOPHILUS PILIPENNIS. n. sp.

Length 1½ lines.

Reddish brown, punctate, and covered with a short yellow pile. Head almost black in front, less narrowed behind, and with the eyes less prominent than in the last species. Antennæ with the club black. Thorax almost square, and somewhat convex, with an indistinct brown patch in the middle. Scutellum large and of a broad triangular form. Elytra of the width of the thorax and truncate, with a basal fascia, large in the middle and smaller towards the humeral angles, of a deep red. The apical portion of the penultimate segment of the abdomen is also red.

287.—CARPOPHILUS OBSCURUS. n. sp.

Length 1 line.

Black, opaque, punctate, and covered with a grey pile which is longest on the sides of the elytra and abdomen. Head very slightly narrowed behind the eyes. Thorax nearly square, a little emarginate in front and rounded at the base. Scutellum large and rounded at the apex. Elytra of the same width as the thorax, and rather obliquely truncate, with a trace of red on each humeral angle. Abdomen with the penultimate segment a little red at the apex. Legs of an uniform red colour.

288.—CARPOPHILUS ATERRIMUS. n. sp.

Length 1½ lines.

Black, opaque, thickly punctate and clothed with short dark pile. Head slightly narrowed behind the eyes. Thorax transverse, emarginate in front, broader behind than at the apex, and rounded at the base. Scutellum large, rounded on the sides and pointed at the apex. Elytra of the width of the base of the thorax, slightly widened behind the humeral angles, and becoming narrowed again towards the apex, which is obliquely truncated, each elytron being cut obliquely from the hinder angle up towards the suture. The legs are brown with the tarsi pale red.

289.—PRIA RUBICUNDA. n. sp.

Length ½ a line.

Entirely of a pale red colour, and punctate. Head small. Thorax broad, flat, emarginate in front, broadly rounded at the anterior angles and on the sides, and truncate at the base, with a large triangular impression on the middle of the disc. Scutellum large, broad, and rounded. Elytra flat, of the same width as the thorax at the base, parallel-sided, rounded at the apex, and nearly covering the entire abdomen.

This very minute insect differs from *P. dulcamara* Illig. the type of the genus in being very flat, in addition to the differences in coloration, marking, and general form. I do not know, however, of any genus of the *Nitidulidæ* to which it makes a nearer approach, and I am unable from the scarcity of specimens, to make in this, as in many other instances, an examination sufficiently minute to enable me to determine positively the exact characters.

290.—SORONIA VARIEGATA. n. sp.

Length 2¼ lines.

Brown, opaque, strongly punctate, with each puncture furnished with a short semi-decumbent seta, the setæ for the most part yellow. Head considerably withdrawn into the thorax. Thorax transverse, deeply emarginate in front, prominent at the anterior angles, rounded, ciliated, broadly margined and reddish on the sides, broadest at the posterior angles, and truncate or slightly bisinuate at the base. Scutellum broadly rounded at the apex. Elytra multi-striate, the striæ punctate and the alternate interstices larger and less interrupted than the others. The insect has a variegated aspect from the disposal of the yellow setæ or setiform scales over the whole surface. On the elytra they are placed so as to give the appearance, under a lens, of three thin yellow fasciæ.

291.—POCADIUS PILISTRIATUS.   n. sp.

Length 1½ lines.

Brown, punctate, and clothed with a long, pale-coloured, decumbent pubescence. Head withdrawn into the thorax up to the eyes which are large but not prominent. Thorax transverse, slightly emarginate in front, much widened behind, slightly bi-emarginate at the base, and more villose at the sides than in the middle. Scutellum broadly triangular, and obtuse at the apex. Elytra not broader than the thorax, parallel-sided, conjointly rounded at the apex, not covering the apex of the abdomen, and covered with close punctate striæ, with the long decumbent pubescence disposed in regular rows along these striæ. There is also a dash of red along the suture and sides of each elytron, and the sides are also strongly ciliated.

292.—NITIDULA CONCOLOR.   n. sp.

Length 1¾ lines.

Black, opaque, punctate, and clothed with short decumbent light coloured pubescence. Head withdrawn into the thorax up to the eyes. Thorax transverse, emarginate in front, rounded on the sides, broad behind, truncate or slightly bi-emarginate at the base, with the posterior angles obtuse. Scutellum broad and rounded at the apex. Elytra not broader than the base of the thorax, very closely and finely punctato-striate and separately rounded at the apex. Abdomen with the last segment large and exposed.

293.—CYCHRAMUS NIGER. n. sp.

Length 2¼ lines.

Black, subnitid, punctate, and moderately pilose. Head small, rather flat, free from pubescence, and sunk to the eyes in the emargination of the thorax. Thorax transverse, emarginate in front, rounded, margined, ciliated, and reddish on the sides, much broader behind than in front, rounded at and behind the posterior angles and nearly truncate at the base. Scutellum with only the apex visible. Elytra about the width of the thorax at the base, and rounded and slightly dehiscent at the apex. Pygidium exposed. Legs red, with the tibiæ strongly ciliated.

294.—IPS POLITUS. n. sp.

Length 2½ lines.

Piceous black, nitid, finely punctate, and subconvex. Head broad and sunk deeply in the thorax. Thorax transverse, deeply emarginate in front, margined on the sides, and broad and truncate at the base. Scutellum rounded at the sides and apex. Elytra scarcely so broad as the thorax at the base, obtusely pointed at the apex, covering all but the very extremity of the abdomen, and marked with a few obsolete striæ. The under side of the body and the legs are punctate and of a pale piceous red.

TROGOSITIDÆ.

295.—LEPERINA MASTERSII. n. sp.

Length 5 lines.

Head black, coarsely punctured, and scaly, with a fascicled tubercle on each side near the eye. Thorax punctate, and covered with black scales excepting on the sides, and two spots near the hinder portion of the disc which are white, and a few cinnamon-coloured scales near the middle, with a smooth raised median line, and two fascicles near the anterior margin protuding forwards. Scutellum rounded behind and covered with cinnamon-coloured scales. Elytra black, coarsely punctured, covered with cinnamon-coloured scales interspersed with white near the scutellum and behind the middle, and spotted with black all over, with three costæ on each elytron, and a small fascicle at the basal end of the second and third, and a larger one near the apex of the first costa. Body beneath black.

L. lacera Pasc., more resemble this species than any other hitherto described, the species are however very different.

296.—LEPERINA GAYNDAHENSIS. n. sp.

Length 7 lines.

Head depressed, black, coarsely punctured, thinly clothed with small yellow scales, with two velvety black spots on the vertex. Thorax with a broad dense mass of yellow scales on the sides, not reaching the anterior angles, with a broad fascia of velvety black between that and the middle, and with the median line coarsely punctured, depressed and almost free of scales. Scutellum transverse and semi-circular. Elytra of a brassy green colour where exposed, covered with large deep punctures, tricostate and marked with numerous velvety black spots, which behind form two narrow fasciæ, with the punctures on the black spots showing under a lens a deep fiery red bottom, while the others are occupied each with a round yellow scale. Body beneath and legs brownish black, punctate and opaque.

297.—LEPERINA BURNETTENSIS. n. sp.

Length 4½ lines.

Head black, coarsely punctured and rather depressed in front, with a fascicle of cinnamon-coloured and white scales on each side near the eyes. Thorax punctate with a dense mass of broad flattened white scales on the sides, a smooth elevated median line, and on each side of it near the anterior margin, three fascicles of long black scales, one in advance of the others and on the margin. Scutellum rounded at the apex. Elytra piceous, punctate, tricostate and thickly clothed with cinnamon-coloured scales, interspersed with a few white ones, with a black fascicle at the base between the second and third costæ, another about the middle between the first and second costæ, a third in a line with the second near the apex, and a few others of smaller size in other positions. The under side of the body is of a dull black and punctate, with a large shallow fovea on the side of each of the abdominal segments.

This species seems to approach the *L. cirrosa* Pasc. It is, however, much less variegated, and the fascicles are much shorter and less numerous.

COLYDIDÆ.

298.—DITOMA COSTATA. n. sp.

Length 1¾ lines.

Brown, flat and opaque. Head punctate, with a shallow depression on each side in front of the eyes. Thorax punctate, and nearly square with the anterior angles advanced, the posterior square and acute, the sides margined and finely serrated, with two costæ on each side curved towards the apex, and joining together on the apical margin, and two others short and ill-defined at the base and near the centre. Elytra scarcely wider than the thorax, parallel-sided, rounded at the apex, punctato-striate,—the alternate interstices three in number, elevated into large costæ,—and marked each with two indistinct patches of deep dull red, one near the humeral angle, extending nearly diagonally to the middle of the elytra and towards the suture, the other an elongate spot, is situated between the middle and the apex.

299.—DERETAPHRUS PASCOEI. n. sp.

Length 5 lines.

Dark brown, subopaque. Head punctate, with a shallow round impression on the anterior margin. Thorax elongate, cordiform, strongly but not densely punctate, and sinuate at the base, with the posterior angles acutely pointed at the termination of a short rather oblique costa, and with the thoracic canal deep at the base, getting shallower towards the interrupted portion which is very short and shallow, and terminating at about one fourth of the length from the apical border. Elytra striated and strongly punctured in the striæ, with the alternate interstices elevated, the third somewhat costiform at the apex and the base, the fifth and seventh costiform throughout, but not large.

300.—BOTHRIDERES MASTERSII.　n. sp.

Length 2½ lines.

Head black and thinly punctured.　Thorax black, subnitid, nearly square, slightly broader in front than behind, finely punctate, slightly advanced at the anterior angles, and strongly and acutely pointed at the posterior, with a large deep and continuous groove in the middle, enclosing a large oblong space of the same level as the rest of the thorax, and with an extension of the groove on the median line at the base, almost to the extremity of the basal lobe.　Elytra red with black suture, nitid, and very finely punctato-striate, with the interstices raised into very narrow costæ.　The legs are piceous.

301.—BOTHRIDERES PASCOEI.　n. sp.

Length 3 lines.

Brown, subopaque.　Head coarsely punctured, with two very faint almost obsolete longitudinal impressions on the forehead. Thorax considerably longer than the width, narrower behind than in front, slightly produced at the anterior angles, square and not pointed at the posterior angles, lobed at the base, and coarsely punctate, with two deep horse-shoe shaped impressions in the middle, one near the base with the open part towards the apex, the other some distance from the apex, with the open part towards the base, the space enclosed by each and intermediate between the two being smooth and almost without punctures. There are also two or more somewhat rugose impressions extending from the hinder impression towards the extremity of the basal lobe.　Elytra punctato-striate, and punctate in rows on the interstices, which are elevated, the alternate ones being strongly costiform.

302.—BOTHRIDERES KREFFTII.　n. sp.

Length 2¾ lines.

This species is very like the last, and both belong evidently to the same group as *B. equinus* and *B. taeniatus* Pasc. It differs from *B.* *Pascoei* in being smaller, in having no trace of frontal impressions, in the thorax being longer and more pointed at the posterior angles, in having the space between the thoracic impressions punctate, in having two short costæ between the hinder thoracic impressions and the base, and in the elytra being of a piceous colour and more strongly costate.

303.—BOTHRIDERES SUTURALIS. n. sp.

Length 1½ lines.

Head and thorax black, subnitid, and coarsely but not densely punctured, the latter nearly square and with a large oblong space in the middle, surrounded except at the base which is open with a broad well marked depression. Elytra red, nitid, finely punctato-striate with the interstices finely costate, and a broad sutural black vitta.

This pretty little species approaches evidently to the group in which Mr. Pascoe places the species *B. vittatus, musivus,* and *merus.*

CUCUJIDÆ.

304.—HECTARTHRUM CYLINDRICUM. *Smith. Col. Brit. Mus.* I., *page* 22.

305.—PROSTOMIS LATICEPS. n. sp.

Length 1½ lines.

Red, nitid. Head broad and triangular, with a deep oblique impression on each side in front of the eyes, and a few punctures on the forehead and vertex. Thorax scarcely so broad as the head, longer than the breadth, finely serrated on the sides, and a little narrowed at the base with the anterior angles acute, the posterior obtuse, and with two crooked interrupted punctured striæ on the disc. Elytra elongate, sub-depressed, not broader than the thorax, parallel-sided, rounded at the apex and punctato-striate, with a black fascia behind the middle.

This species has not got the advanced mandibles of *P. mandibularis,* nor do the antennæ agree with the generic characters given by Lacordaire. The first joint is thick and longer than the others, the second a little smaller, the third and following joints up to the 9th are short and moniliform, and the last three form a compact club of which the basal joint is the largest.

306.—Ipsaphes nitidulus. n. sp.

Length 2 lines.

Brownish red, nitid and punctate   Antennæ with the third joint scarcely longer than the fourth.   Head with a slight longitudinal impression on the forehead.   Thorax transversely quadrate, emarginate on the anterior border, and truncate at the base, with the angles produced into minute teeth, and with the sides a little rounded behind the anterior angles.   Scutellum transverse and slightly rounded behind.   Elytra strongly punctato-striate, with a sublateral carina.

PLACONOTUS. n. gen.

Like the genus *Platisus* Erichs ; but differs entirely in the character of the antennæ.   The body is very flat.   The antennæ are long and filiform, the first joint thick, the second, third, and fourth joints of nearly equal size, slighter and shorter than the first, the remainder long and slight.   The first article of the tarsi is of normal length.

307.—Placonotus longicornis. n. sp.

Length 1 line.

Brownish red, nitid, finely punctate.   Antennæ of the length of the body.   Thorax nearly square, with a small tooth at each angle, and a sublateral line.   Scutellum transverse and triangular. Elytra of a pale red, and very finely punctato-striate, with a broad longitudinal depression in the middle of each elytron.

308.—Brontes nigricans, Pasc. *Journ. of Ent., Vol.* I., *page* 321.

309.—Silvanus castaneus. n. sp.

Length 1½ lines.

Of an opaque chesnut colour, mixed slightly with black in

some places, and densely punctate. Thorax much longer than the breadth, flattened in the middle and finely serrated on the sides. Elytra closely punctato-striate, with a depression behind the scutellum.

310.—OMMA STANLEYI, Newm. *Ann. Nat. Hist.* III., *page* 303.

311.—OMMA MASTERSII. n. sp.

Length 4 lines.

Black, opaque, densely punctate and closely covered with scales. Head flat, sprinkled with white scales, and affixed to the thorax by a large neck. Thorax about the width of the head, broader than the length, much rounded at the anterior angles, deeply and largely bi-foveated in front, truncate at the base, and sprinkled with white scales over the basal portion. Scutellum transverse, truncate, and covered with white scales. Elytra broader than the thorax at the base, gradually enlarging towards the apex, separately rounded at the apex, coarsely striato-punctate, and marked with a humeral spot, a median wavy fascia, a lateral vitta not reaching the base, and a sutural vitta confined to the apical third, composed of white scales. The under side of the body and the legs are closely covered with white scales.

LATHRIDIDÆ.

312.—CORTICARIA POLITA. n. sp.

Length ⅔ of a line.

Greenish-black, nitid, convex. Head punctate. Antennæ with the base red, and club black. Thorax ˙ nearly square, not broader than the head, a little rounded on the sides, covered with large punctures, with two foveæ near the base, and four setigerous points on the sides, one of them forming the posterior angle. Elytra broader than the thorax and rounded at the apex.

MYCETOPHAGIDÆ.

313.—TRIPHYLLUS FASCIATUS.   n. sp.

Length 1¼ lines.

Black, subnitid, punctate, very hairy and of an oval form. Thorax transverse and margined on the sides.   Elytra red with a broad black median fascia, and strongly striato-punctate.

314.—DIPLOCOELUS OVATUS.   n. sp.

Length ⅔ of a line.

Reddish brown, subnitid, covered with long yellow hair, and of a short oval convex form.   Thorax transverse, margined at the sides, and with two striæ near the sides.   Elytra strongly striato-punctate.

DERMESTIDÆ.

315.—DERMESTES MURINUS, Linn.   *Faun Suec., page* 144.—Erichs. *Nat. Ins.* III., *page* 429.— Bouché. *Nat. Ins., page* 189 ; *catta,* Pang.   *Naturf.* 24, 10, *t.* 1., *f.* 12.— Herbst. *Käf.* IV., *page* 123. *t.* 40, *f.* 4 ; *nebulosus* De Geer. *Ins.* IV., *page* 197; *rosciventris* Casteln. *Hist. Nat.* II., *page* 34.

316.—MEGATOMA APICALIS.   n. sp.

Length 2½ lines.

Black, subnitid, punctate, and clothed with black hair. Thorax transverse, convex, with the base triangularly produced. Elytra of the width of the base of the thorax, with the apex dehiscent, separately rounded, not covering the pygidium and of a piceous hue.   Under side of the body punctate and covered with a decumbent ashen pile.   The legs and antennæ are of a piceous red.

317.—ANTHRENUS NIGRICANS. n. sp.

Length 1½ lines.

Black, punctate, with the head and thorax sprinkled with a whitish pubescence, and with two interrupted fasciæ of the same on the elytra. The under surface of the abdomen and the pygidium are densely covered with orange coloured pubescence.

318.—CRYPTORHOPALUM OBSCURUM. n. sp.

Length 1½ lines.

Black, opaque, punctate and covered with a short decumbent yellowish pile. Basal lobe of thorax broadly rounded. Elytra with two distinct lateral striæ, and faint traces of others over the rest of their surface.

I may not be correct in placing this insect in the genus *Cryptorhopalum*; it will probably be found on an examination more minute than I have been able to give it to form a new genus.

319.—TRINODES PUNCTIPENNIS. n. sp.

Length 1⅓ lines.

Black, subnitid, punctate and very hairy. Thorax strongly bisinuate at the base, with the median lobe slightly emarginate. Scutellum smooth, triangular and somewhat convex. Elytra coarsely punctured in irregular rows. Legs of a piceous red colour.

320.—TRINODES GLOBOSUS. n. sp.

Length ⅔ of a line.

Besides the great difference in size this species differs from the last in being very short and convex, in having the base of the thorax less bisinuated, and in having the scutellum nearly round.

BYRRHIDÆ.

321.—MICROCHÆTES FASCICULARIS. n. sp.

Length 2 lines.

Black, punctate, and covered all over with erect scales.

Thorax transversely impressed in the middle, with a transverse series of five fascicles immediately behind the impression. Elytra coarsely punctato-striate, with four rows of lengthened fascicles on each, one on the third interstice consisting of five or six fascicles, the second on the fifth interstice of four fascicles, the third on the seventh of three, and the fourth on the ninth of two. There are a few brown scales mixed with the black in some places.

This species is larger than *M. sphaericus* or *Australis*, and more regularly striated on the elytra, while in *M. scoparia* the striation of the elytra is much finer than in any of them.

322.—MICROCHÆTES COSTATUS. n. sp.

Length 1 line.

Black and scaly. Thorax not produced in front over the head as in the other species of the genus, but receiving the head in the emargination of the anterior border. Elytra very finely striate-punctate, with five narrow costæ on each, and with the intervals quite flat.

This ought probably to constitute a new genus.

323.—LIMNICHUS FRONTALIS. n. sp.

Length ⅔ of a line.

Black, opaque, finely punctate. Head with a large triangular impression on the forehead. Thorax very broad, scarcely emarginate in front, and widening much towards the base, which is broadly rounded. The elytra are very finely punctato-striate.

GEORYSSIDÆ.

324.—GEORYSSUS KINGII. n. sp.

Length ½ a line.

Black, opaque. Thorax rough, granulose, rounded and rather produced over the head in front, and a very little emarginate at the apex, with the median line broadly marked, and the base broadly rounded. Elytra shortly ovate and convex, with four

large costæ on each, and with the intervals occupied by transversal quadrangular foveæ.

The sculpture of this insect is very different from that of *G.
Australis* King ; the only Australian species hitherto described.

HETEROCERIDÆ.

325.—HETEROCERUS MASTERSII. n. sp.

Length 1⅔ lines.

Black, finely punctate and pubescent. Thorax rounded at the
base, with a broad lateral testaceous border. Elytra obsoletely
striated and of a testaceous colour, with a number of black spots,
which form themselves into apparently very interrupted fasciæ,
one basal, one median, and the other apical. The legs and mandibles are also testaceous.

LUCANIDÆ.

326.—LAMPRIMA KREFFTII. n. sp.

Length 9 lines, mandibles included.

Though much smaller, this species most resembles *L. Latreillii*
MacLeay ; it is, however, evidently a different species. The most
evident points of difference are :—the thorax of this species is
more sparsely punctured, but the punctures are much larger, the
scutellum is more triangular and is minutely emarginate at the
tip, and the elytra are less rugose and have a distinct stria near
the suture on each side. The pubescence also on the pygidium
seems to be of a much darker colour.

> ### 327.—FIGULUS REGULARIS, Westw. *Ent. Mag. V.,*
> *page 263.*

> ### 328.—FIGULUS LILLIPUTANUS, Westw. *Trans. Ent.
> Soc.,* 2 *Ser.* III., *page* 219. *t.* 12, *f.* 5 ;
> *clivinoides Thoms. Ann. Fr., page* 432.

329.—AULACOCYCLUS KAUPII. n. sp.

Length 13 lines.

Like *A. edentulus* McLeay. Differs from it in having the

labrum more emarginate, the frontal horn less arched forwards, less emarginate at the extremity, and more sulcate on the back, in the lateral fovea on the thorax being more uninterruptedly semilunar, and in having the striæ on the elytra more profound.

I name the species after Dr. J. J. Kaup, of Darmstadt, who has made the *Passalidæ* the object of his study for some time.

330.—TÆNIOCERUS MASTERSII.   n. sp.

Length 9 lines.

Black, nitid, and of an elongate form, frontal horn short, obtuse, and slightly sulcate on the back. Thorax transverse, rather rounded behind and deeply impressed on the median line, with the lateral foveæ shorter, less semilunar and nearer the posterior angles than in the species last described.   Elytra punctato-striate, the punctures indistinct in the first three striæ from the suture.   Club of the antennæ large.

331.—MASTOCHILUS AUSTRALASICUS, Perch. *Suppl.* I., *page 6, t. 77, f. 2.*

332.—MASTOCHILUS POLYPHYLLUS, McLeay. *King's Survey* II., *page 439.*—Burm. *Handb.* V., *page 469.* *Hexaphyllus* Boisd. *Voy. Astrol. Col., page 241.* *Sexdentatus* Perch. *Mon., page 28, t. 2, f. 5.*

333.—MASTOCHILUS NITIDULUS.   n. sp.

Length 15 lines.

Like *M. dilatatus* Dalm. Head very rugose, with the frontal ridge rather long, and from the extremity of it a small ridge extending obliquely to the basal angles of the labrum, in the middle of these oblique ridges there is a small tubercle, and a low transverse ridge unites these two tubercles, thus enclosing a small triangular space in front of the frontal ridge. In these particulars the sculpture of the head in this species differs widely from that of *M. dilatatus*, where the tubercles are larger, and there are two large transverse ridges, one on the forehead, the other a shorter one near the front. The thorax has the faintest possible trace of the median line in the centre, has the lateral foveæ deep, round, near the posterior angles, and without other punctures near them as in *M. dilatatus*, and has a shallow fovea containing two or three punctures near the anterior angles. The elytra are of the same form and sculpture as *M. dilatatus*. The fore tibiæ have their external surface punctured on the apical half only.

334.—MASTOCHILUS PUNCTICOLLIS. n. sp.

Length 15 lines.

The sculpture of the head is much the same as in the last species, but the space in front of the frontal ridge is shorter and more perpendicular. In addition the whole insect is less nitid, there is no trace of median line on the thorax, the foveæ are more filled with small punctures, and there is an accumulation of punctures near the anterior angles. The fore tibiæ also have their external surface punctured almost to their base, and their external teeth are very obtuse.

SCARABÆIDÆ.

Sub-family COPRIDÆ.

CANTHONOSOMA. n. gen.

Body broadly ovate, convex. Head flat, transverse, broadest and angular in the middle, and broadly rounded behind and in front, with a small semi-circular emargination between two small tubercles at the apex. Eyes oval above and free on the posterior edge. Mentum and labial palpi resembling those of *Cephalodes-mius*. Thorax transverse, convex, flattened on the sides, rounded behind and emarginate in front. Elytra as broad as the length, sub-convex, much narrowed at the apex with the marginal epipleuræ deep. Pygidium perpendicular, sub-triangular and rounded at the apex. Legs moderately stout; the anterior without tarsi, with the tibiæ slightly arcuated and furnished with a small tooth on the outside near the apex, two short obtuse teeth at the exterior apical angle and a strong acute spur at the inner

L

angle; the intermediate and posterior, long, with the tibiæ arcuated, quadrangular and ciliated, and with two acute spurs at the apex of the intermediate and one at the apex of the posterior tibiæ.

The absence of the anterior tarsi separates this genus from the other Australian genera of long legged *Copridæ*. According to Lacordaire's arrangement the genus would be included in the small group which he names *Deltochilides*. I have two species in my cabinet, one from Rockhampton, the other from the Pine Mountains, Queensland.

335.—CANTHONOSOMA MASTERSII. n. sp.

Length 8 lines.

Black, subopaque. Head and thorax finely punctate, the latter with the anterior angles acute, the posterior rounded, the sides also rounded with a shallow fovea at the broadest part, and with a very faint indication near the base of the median line, and on each side of it a smooth tuberosity. Elytra scarcely so broad as the thorax, thinly and finely punctate—each puncture furnished with a short semi-decumbent yellow seta—and costate, the costæ broad, flat, obsolete looking, and six in number on each elytron. The pygidium and under surface of the body smooth.

This insect is common throughout the Northern Districts, and is invariably found on Wallaby dung.

336.—CEPHALODESMIUS QUADRIDENS. n. sp.

Length 3 lines.

Black, opaque, punctate. Head densely punctured on the forehead, and quite smooth and nitid in front, with four strong acute and slightly recurved teeth on the margin of the clypeus. Thorax, with the punctures of an oval shape and setigerous, the median line distinct especially towards the base, and a fovea near each lateral margin. Elytra striate, punctate,—the punctures setigerous,—and subcostate, the costæ broad and depressed. Under surface nitid. Legs piceous, with the tibiæ profoundly punctured.

337.—TEMNOPLECTRON TIBIALE. n. sp.

Length 2¼ lines.

Black, nitid, convex and nearly round. Head almost vertical, and finely punctate with the clypeus angularly emarginated at the apex. Thorax transverse, very finely punctate, emarginate in front, slightly rounded behind, broader at the middle than at the apex, and parallel sided from the middle to the base. Elytra of the width of the thorax at the base, rounded at the apex, not longer than the breadth, very finely punctate and marked with seven very fine striæ—the external one abbreviated,—on each elytron. The fore tibiæ are almost rectangularly bent inwards near the apex, and at the external angle formed by the bend there is a strong tooth, with two small teeth above. The four posterior tibiæ are slightly curved, enlarged towards the apex, and laterally compressed. The antennæ and palpi are red.

MERODONTUS. n. gen.

Body broadly ovate. Head flat, transverse, rounded behind, and emarginate in front, with three minute teeth along the anterior edge of the clypeus on each side of the emargination. Eyes only visible through a narrow slit in the hind margin of the head. Mentum narrow and emarginate. Labial palpi with the last joint obconic, and about half the length of the penultimate. Thorax transverse, subconvex, slightly rounded at the base, and emarginate in front, with the anterior angles much enlarged. Elytra not covering the pygidium, and flat on the back, with the sides and apex deep and vertical. Anterior legs rather short, with the tibiæ wide at the apex and minutely tridentate, and the tarsi very short; intermediate, long with the tibiæ strongly tridentate; posterior, very long, with the thighs thick and furnished in one sex with a large acute spur near the apex on the under side, and with the tibiæ very much arcuated. Pygidium perpendicular.

The insect on which this genus is formed has a strong general resemblance to the genus *Sisyphus*, the deep epipleuræ of the elytra clearly indicate, however, its true position to be with the group named *Minthophilidæ* by Lacordaire.

338.—MERODONTUS CALCARATUS. n. sp.

Length 4 lines.

Piceous brown, opaque, densely punctate, and clothed with short semi-erect pale coloured hairs. Head with a slight convexity in the middle, and a short slightly oblique transverse ridge on each side in front of the eyes. Thorax with two abbreviated ridges near the median line composed each of two elongated tubercles, extending from the apical margin to the middle, with six small conical tubercles placed at intervals across the thorax in a semicircular form, with another tubercle near the posterior angles, and with the anterior angles large, flat, and obliquely truncate, the hairs on the anterior angles, and on the tubercles being longer than on the rest of the surface, and more curled at their extremities. Elytra punctato-striate,—the striæ consisting each of two fine lines with flat interstices, the alternate interstices red, the others having two or more small tubercles,— strongly costate at the lateral margin of the flat dorsal surface,— the costa not reaching the suture, and terminating in a reddish tubercle,—and furnished with a fine costa near the lateral margin of the epipleura. Pygidium convex and tuberculate, the tubercles small and taking a V shaped position. Under surface piceous black, opaque, punctate and somewhat scaly. Legs strongly punctate and ciliated with short reddish hair.

I have seen specimens of this insect from Moreton Bay, and I believe it is pretty generally distributed over the whole of the Northern parts of New Holland. It is found, as is also the case with the last described species *Temnoplectron tibiale*, in human excrement.

339.— ONTHOPHAGUS GRANULATUS, Bohem. *Res. Eugenie* 1858, *page* 48.

340.— ONTHOPHAGUS QUADRIPUSTULATUS, Fab. *Spec. Ins.* I., *page* 31,—Oliv. *Ent.* I. 3., *page* 175, *t.* 15. *f.* 141.—*Montrouz. Ann. Soc. Agr. Lyon.* VII., 1857, I., *page* 22.—Harold. *Col. Heft.* II., *page* 32.

In the "Catalogus Coleopterorum" of Dr. Gemminger and B. von Harold, I find that the species described by me some years ago under the name of *O. rubrimaculatus*, is put down as a synonym of this species. This is incorrect, I have both species, there is a marked distinction between them as regards both size and sculpture.

I may here mention two other errors in that publication. My *Onthophagus furcatus* which is put down in the Catalogue alluded to as a synonym of *O. auritus* Erich., is as widely different from that insect as one species of *Onthophagus* can possibly be from another, and my *O. laminatus*, put down as a synonym of *O. capella* Kirby, is very distinct from that species.

341.—ONTHOPHAGUS CUNICULUS, MacL., W. *Trans. Soc. N. S. Wales*, 1864, I., *page* 123.

In my description of this species, I made the mistake of describing as the male a female with the thoracic tubercles more than usually large. The male has the thorax more retuse in front than the female, and has a strong triangular horn in place of the two tubercles of the female.

342.—ONTHOPHAGUS DIVARICATUS. n. sp.

Length 4 lines.

Black, subopaque. Head, in the male with a transverse raised line between the eyes, and two long rather slender and slightly curved horns emanating from each extremity of that line, extending in a direction upwards and outwards, and obtusely pointed at their apex; in the female with the frontal transverse raised line rising on each side from the upper angle of the eye, and without horn or tubercle; in both sexes, with the clypeus rugosely and transversely punctured, with its suture raised and semicircular, and with the space behind the frontal transverse ridge smooth and nitid. Thorax transverse, convex, opaque, densely punctate, and presenting a granulated appearance, with the sides bulged out behind the middle, with a fovea near the bulge, with the posterior angles emarginate, and without any trace of the median line. The thorax of the male is less coarsely punctate than that of the female, is furnished with a strong laterally compressed horn, and is a little retuse and nearly smooth in front. Elytra having a shagreen appearance, with seven fine transverse punctured striæ on each elytron, and with the interstices flat. Legs ciliated with red hair. Antennæ, palpi, and tarsi, red.

The female of this species very closely resembles the female of O. *granulatus*, described by me from Port Denison in 1864, which species, I see by the " Catalogus Coleopterum," has had its name changed by Von Harold to *consentaneus*, I presume because the name *granulatus* had been previously applied by Boheman to another species. I am inclined to believe that in the case of that species I made the same mistake as in O. *cuniculus*, and described two females only and not the male. The difference between the females of O. *consentaneus* and the one now described consists chiefly in *consentaneus* having the median line of the thorax visible, in having the elytra much more roughly marked, in having the under side of the body more hirsute, and in having the space on the head behind the frontal transverse ridge, punctate.

343.—ONTHOPHAGUS RUBICUNDULUS. n. sp.

Length 1½ lines.

Black, opaque, punctate, and marked on the margin of the head and clypeus, the sides and base of the thorax, and over all parts of the elytra with indistinctly defined deep red spots. Head transverse, bisinuate slightly on the sides, emarginate at the apex, and of a rather brassy hue with two long slightly curved horns on the vertex, extending upwards and outwards, and obtusely pointed, in the males, and in the females two small tubercles occupying the place of these horns. Thorax transverse, emarginate in front, rounded behind, considerably bulged out on the sides, foveate at the broadest part, and shallowly variolous-punctate on the surface, with the punctures each furnished with a decumbent yellow seta. The male has in addition an obtuse protuberance near the anterior part of the thorax, and a rather deep excavation on each side, of a slightly metallic hue. Elytra finely striate, with the interstices flat and having in the middle of each a row of punctures bearing yellow setæ. The legs are of a piceous red.

344.—Onthophagus perpilosus. n. sp.

Length 2¼ lines.

Black, subnitid, strongly punctate, and densely pilose. Head with two strong transverse ridges, with the margin of the clypeus recurved and the apex emarginate. Thorax with a slight coppery gloss on the anterior part, a faint fovea near the sides, and a faint transverse ridge in the centre near the apex. Elytra striate with the interstices broad, somewhat costiform and punctured in rows. Body beneath black, nitid, punctate, and moderately hairy, the hair white. Legs piceous.

The two specimens of this insect in the collection present some differences which may be sexual. In one, the teeth of the fore tibiæ are more large and pointed, the transverse ridge on the thorax more distinct, and the emargination of the clypeus larger than in the other.

345.—Onthophagus incornutus. n. sp.

Length 2⅓ lines.

Black, subnitid, very finely punctate. Head, in the male rather rugosely punctured with the forehead depressed, with a slightly elevated curved transverse swelling behind, and with the apex of the clypeus almost truncate ; in the female with the forehead flat and slightly rugose, and with the clypeus slightly emarginate. Thorax in the female of a coppery hue. Elytra opaque and finely punctato-striate, with the interstices broad and flat, and exhibiting under a lens traces of red spots at the humeral angles and apex. Legs piceous, with the dentations of the fore tibiæ very strong in the female. Antennæ pale red.

346.—Onthophagus Mastersii. n. sp.

Length 4 lines.

Black, nitid, densely punctate. Head, in the male large, rounded in front with the margin reflexed, and armed with two long rather slender parallel horns on the forehead and with a short broad truncate plate between them ; in the female with two transverse ridges and the apex of the clypeus almost truncate. Thorax, in the male convex and strongly punctured behind, very retuse and smooth in front, and with the elevated portion more advanced on the median line than at the sides, in the female without tuberosity of any kind, and without any trace of the median line. Elytra punctato-striate with the interstices broad, moderately flat and punctate. Body beneath somewhat hairy. Legs black. Antennæ red.

347.—ONTHOPHAGUS CAPELLA, Kirby. *Trans. Linn.*
Soc. XII., *page* 398.

348.—ONTHOPHAGUS DESECTUS. n. sp.

Length 6 lines.

Black, nitid, and densely punctate. Head with two transverse ridges in both sexes, the clypeus a little more pointed and reflexed in the male than in the female. Thorax, in the male elevated into a tubercle on the median line behind, and presenting from that point to the anterior margin an oblique flat surface as if cut down, on each side of the central tubercle and a little in front there is a small protuberance also cut through in the same way, and on the oblique anterior surface the median line is finely carinated ; in the female there is a tubercle in the middle of the median line, a slight depression in front, and the median line itself is carinated finely in front and slightly impressed behind. Elytra with seven fine smooth lightly punctate striæ on each, with the interstices broad, flat, and smooth. Anterior tibiæ slight and flatly toothed in the male.

This species has a wide range and is not by any means uncommon in many parts of New South Wales. I cannot find however that it has ever been described.

349.—ONTHOPHAGUS QUINQUETUBERCULATUS. n. sp.

Length 8 lines.

Black, subnitid, finely and densely punctate. Head transversely punctate, with the clypeus rounded and reflexed at the apex in the female, and almost truncate and more reflexed in the male, and with the frontal transverse ridge almost obsolete in the middle, and pointed at each end in the male, while in the female it is of uniform size throughout. Thorax in the male, with a prominent advanced tubercle in front and in the middle, and two small tubercles on each side placed along the line of the retuse anterior front of the thorax, which is perpendicular and considerably behind the advanced central tuberosity, which is slightly emarginated by the median line ; in the female, with the advanced central prominence also submarginate, but very broad, and with only one small tubercle between it and the sides. Elytra with eight fine striæ on each, and with the interstices broad, flat, and very finely punctate. Body beneath clothed with long fulvous hair.

350.—ONTHOPHAGUS INERMIS. n. sp.

Length 3½ lines.

Black, nitid. Head very finely and transversely punctate on the forehead, and densely and rugosely on the clypeus, which is broadly rounded and moderately reflexed along the margin. Thorax quite smooth and without any mark excepting the lateral fovea. Elytra punctato-striate, with the interstices almost quite flat and smooth. Tibiæ piceous.

This species is not unlike *O. muticus* described by me from Port Denison, in the transactions of the Entomological Society of New South Wales, for the year 1864. It differs from it in being smaller, more brilliant, smoother on the thorax, and more deeply striated on the elytra. The only difference I can perceive between the sexes, both in this species and in *O. muticus*, is in the much more strongly developed teeth on the exterior of the anterior tibiæ in the female.

351.—APHODIUS GEMINATUS. n. sp.

Length 1⅓ lines.

Black, subnitid. Head coarsely punctate, rugose, toothed and emarginate in front. Thorax densely punctate, transverse, bisinuate at the sides, not broader behind than in front, and rounded at the base, with a broad depression near each side, and a deep broad transverse depression behind the middle, which extends to the base in the centre. Elytra finely striate in pairs, with the interstices flat and smooth, and with a few minute elongate tubercles on the alternate interstices. Legs piceous.

In one specimen before me there are no traces of tubercles on the elytra.

352.—Aphodius lividus, Oliv. *Ent.*, 1. 3, *page* 86, *t.* 26, *f.* 222, *&c. &c.*
cincticulus Hope, *Trans. Ent. Soc. Lond.* IV., 1847, *page* 284, *S. Australia.*
spilopterus Germ. *Linn. Ent.* III., 1848, *page* 189, *S. Australia.*

353.—Ammœcius obscurus. n. sp.

Length 2 lines.

Brownish black, opaque, densely punctate, and somewhat squamose. Head slightly but broadly emarginate in front. Thorax transverse, almost truncate at the apex, nearly parallel and bisinuated at the sides, and slightly rounded at the base, with the anterior angles large and slightly reflexed, and the posterior obtusely pointed backwards. Elytra striate, with the striæ large and filled with shallow punctures, and with the interstices narrow and costiform.

In one specimen of this insect before me, the thorax has on each side two abbreviated transverse depressions.

354.—Ammœcius crenatipennis. n. sp.

Length 1½ lines.

Black, densely punctate, subopaque. Head broadly rounded and emarginate in front. Thorax transverse, and rounded at the base, with the sides subparallel, the anterior angles somewhat dilated, the posterior not prominent, and the median line slightly marked at the base. Elytra punctato-striate, the punctures large and giving a crenulated appearance to the costiform interstices.

355.—Ammœcius semicornutus. n. sp.

Length 1½ lines.

Black, subnitid. Head marked with very fine longitudinal

punctures, with the clypeus deeply emarginate and of a reddish colour, and with the back of the head on each side furnished with a very minute tubercle or tuberosity, which is extended in a raised line to the lateral border. Thorax transverse, and not densely punctate, with a large shallow depression near the sides, with all the angles rounded, and with the median line distinct at the base. Elytra strongly punctato-striate, with the interstices elevated and smooth. Legs piceous.

Salvia ?

356.—AMMŒCIUS NITIDICOLLIS. n. sp.

Length 1½ lines.

Piceous black, nitid. Head almost smooth, reddish in front and emarginate. Thorax transverse, almost smooth in front and thinly punctate behind, with the anterior angles rounded, the posterior square, and the base very slightly rounded. Elytra strongly punctato-striate, with the interstices broad, elevated and smooth.

357.—BOLBOCERAS GAYNDAHENSE. n. sp.

Length 5½ lines.

Reddish chesnut, nitid. Head with a small short transverse ridge in the centre of the occiput elevated at each end into a minute tubercle, and with the forehead depressed, punctate, and enclosed on the sides and in front by a narrow ridge, which is elevated in the middle in front into a minute horn, from which horn two elevated lines extend in a diagonal direction to an anterior short transverse ridge. Thorax sparsely punctate, with a large round excavation in the middle of the front. Elytra punctato-striate, the punctures minute. Body beneath densely fulvous pilose.

There are only two specimens in the collection, both evidently females.

358.—TROX SULCARINATUS, McLeay, W. *Trans. Ent. Soc. N. S. Wales,* I., *page* 128.

359.—Trox squamosus. n. sp.

Length 8 lines.

Black, opaque, densely squamose. Head punctate, bitubercu-late and clothed beneath with red hair. Thorax transverse, punctate, and interruptedly six-costate,—the costa on each side of the median line diverging in the middle and not reaching the base, the next entire, close to the last and merging in it in the middle, and the third near the posterior angles, thick, interrupted, and not extending to the apex—with the lateral margin trituber-culate, one tubercle at the posterior angles, the other two forming sinuations on the sides behind the anterior angles. Scutellum lanciform. Elytra much broader than the thorax, acute at the humeral angle, gradually widening behind, broadly rounded and minutely emarginate at the apex, and roughly punctured in irregular rows, with the alternate interstices furnished with large elongated tubercles, and with the first of these alternate inter-stices costate at the base. Fore tibiæ unarmed.

360.—Trox salebrosus. n. sp.

Length 3¾ lines.

Black, densely punctate and lightly squamose. Head bitu-berculate and clothed beneath with yellowish hair. Thorax sculptured much in the same manner as in the preceding species *(T. squamosus)*, the central costæ however being less divergent in the middle, those next to them being more distant and quite interrupted in the middle, and the two tubercular sinuations of the anterior part of the lateral margin being much more slight. Elytra very roughly and coarsely punctured in rows, with small elongated tubercles on the interstices, the tubercles on the alter-nate interstices being the largest. Fore tibiæ with a small tooth in the middle of the external margin and a large truncate lami-nated one of a reddish colour at the apex.

361.—Trox semicostatus. n. sp.

Length 5 lines.

Black, punctate, slightly squamose. Head bituberculate with

reddish yellow hair beneath. Thorax more sinuate on the sides
than in the last species, and with the costæ more elevated—the
median costæ almost reaching the base. Scutellum broad and
lanciform. Elytra punctured in rows with the alternate interstices strongly costate at the base and towards the apex interrupted and formed of tubercles, and with the others marked with
elongate tubercles of smaller size ; there is also near the humeral
angle a rugose tuberosity and a short elevated costiform line.
Fore tibiæ obtusely toothed in the middle and roundly and not
prominently at the apex.

362.—LIPAROCHRUS SCULPTILIS, Westw. *Trans. Ent.*
Soc. Lond. 2nd Ser. II., page 70.

Sub-family MELOLONTHIDÆ.

363.—PHYLLOTOCUS NAVICULARIS, Blanch. *Cat. Coll.*
Ent., 1850, *page* 97.

364.—PHYLLOTOCUS SERICEUS. n. sp.

Length $3\frac{1}{3}$ lines.

Head black, finely punctate. Thorax red very sparingly
punctured and fringed with erect black hairs. Elytra red with
a narrow black lateral margin, silky, separately rounded and
dehiscent at the apex, strongly striate with the interstices
rounded, and furnished along the lateral border, the suture, and
the interstices between the striæ with erect black somewhat
distant hairs. Body beneath and legs black and punctate, with
the exception of the prothorax and anterior coxæ which are red.

Of all the species of *Phyllotocus* hitherto described, this one
most resembles *P. Australis* Boisd.

365.—PHYLLOTOCUS VARIICOLLIS. n. sp.

Length $3\frac{1}{2}$ lines.

Head black and punctate. Thorax punctate, thinly fringed
with black hairs, entirely black in the male and red with a cen-

tral black vitta in the female.   Elytra dark red with the suture
and lateral margin behind black, subnitid, rather rugosely punctate, and deeply striate with the interstices much rounded.
Under surface and legs black, excepting the lower surface of the
fore thighs and tibiæ and the ungues of the tarsi which are red.

The male of this species resembles my *P. marginipennis*, and
the female my *P. marginatus*, it differs from both in having the
head and thorax more coarsely punctured, and in the elytra
being punctate and entirely destitute of any silky gloss or velvety
texture.

366.—DIPHUCEPHALA AUROLIMBATA, Blanch. *Cat. Coll.*
Ent. 1850, *page 99.*

367.—MÆCHIDIUS VARIOLOSUS.   n. sp.

Length 4½ lines.

Piceous brown, subnitid, densely punctate, and clothed with
erect soft reddish brown hairs.   Head slightly emarginate and
reflexed in front.   Thorax with the anterior angles slightly advanced and rounded, and the posterior acute.   Elytra closely
covered with series of elongate oval variolous punctures each
with an erect soft hair at its anterior margin, and with the alternate interstices a very little broader than the others.   Legs
piceous red.

368.—MÆCHIDIUS OBSCURUS.   n. sp.

Length 5½ lines.

Black, opaque, densely variolous, punctate and setose.   Head
much reflexed and slightly emarginate in front.   Thorax roughly
granulose, and shallowly bifoveate near the sides, with the
median line lightly marked and the sides near the posterior angles
obliquely narrowed, and very slightly emarginate.   Elytra closely
covered with regular rows of elongate variolous punctures, each
with a very short decumbent yellow seta at its anterior margin.
Body beneath piceous black, subnitid and densely punctate.

369.—MÆCHIDIUS RUGOSICOLLIS.   n. sp.

Length 4½ lines.

Brownish black, opaque, densely variolous, punctate and

setose. Head slightly bisinuate at the sides, and emarginate in the middle. Thorax rough, bifoveate near the sides and profoundly emarginate before the posterior angles. Elytra with the punctures less elongate and less sharply defined than in *M. obscurus*, and with elevations at the base, which extend in the form of obsolete costæ almost to the apex. Pygidium deeply emarginate. Body beneath subnitid.

370.—MÆCHIDIUS PARVULUS. n. sp.

Length 3 lines.

Piceous brown, subnitid, densely variolous-punctate and setose. Head bisinuate and reflexed at the sides, and deeply emarginate in the middle. Thorax narrowed and rounded towards the posterior angles which are sharp. Elytra closely covered with regular rows of roundish variolous punctures, each with a decumbent yellow seta the length of itself, and a stria between every two rows of punctures.

371.—LIPARETRUS FULVOHIRTUS. n. sp.

Length 4 lines.

Head and thorax black, densely punctate, and closely covered with erect soft pale red hairs ; the latter with the punctures coarser than those of the head, and with the median line visible. Elytra red, except on the basal margin, separately rounded and somewhat dehiscent at the apex, thinly clothed with erect hairs, and coarsely and irregularly punctate, with the three geminate striæ rather indistinct. Pygidium and abdominal segments black, punctate and thinly clothed with long light coloured hairs. The under side of the thorax densely clothed with hairs of the same pallid hue. Legs red.

372.—LIPARETRUS SERICEUS. n. sp.

Length 3¾ lines.

Head and thorax black, pruinose, very finely punctate. Elytra dark red with the base and lateral margin black, broadly rounded or almost obliquely truncate at the apex, and irregularly punctate, with four rather well defined geminate punctate striæ on each elytron. Pygidium black, and finely punctate. Body beneath black, and moderately cinereo-pilose. Legs piceous black.

373.—LIPARETRUS PILOSUS. n. sp.

Length 3¼ lines.

Head black, punctate, densely pilose with the clypeus of the male broad, reflexed, and acutely pointed outwards at the angles. Thorax also black, densely punctate, and thickly covered with long erect soft light brown hairs, with the median line marked at the base. Elytra very dark red with the suture, base, and lateral margins black, coarsely and irregularly punctate, clothed with erect hairs not quite so thick or so long as on the thorax, and separately rounded at the apex. Pygidium and under surface of body black, punctate and densely clothed with a long ashen pubescence. Tarsi reddish.

374.—LIPARETRUS ATRICEPS, MacL., W. *Trans. Ent. Soc. N. S. Wales*, I, *page* 128.

375.—LIPARETRUS PALLIDUS. n. sp.

Length 2¾ lines.

Of a pale red, nitid, punctate, and without pubescence on the upper surface. Head black, with the clypeus broad, short, rounded at the angles, nearly truncate, and of a reddish colour. Thorax with the median line lightly marked. Elytra irregularly punctate each with three tolerably distinct geminate striæ, and broadly rounded at the apex. Pygidium finely punctate and clothed with short hair. Under side of body rather thinly clothed with reddish hairs.

376.—LIPARETRUS FLAVOPILOSUS. n. sp.

Length 3¼ lines.

Head and thorax black, finely punctate, and covered with long yellowish somewhat recumbent hair; the former with the clypeus acute and reflexed at the angles, and very slightly emarginate in the middle ; the latter with the median line distinct near the base. Elytra testaceous red with the basal margin black, subnitid, thinly clothed with hair, and irregularly punctate, with the geminate striæ rather indistinct. Pygidium and under surface of body black and flavo-pilose.

377.—LIPARETRUS RUFIVENTRIS. n. sp.

Length 3½ lines.

Head and thorax black, punctate, and covered with long reddish yellow somewhat recumbent hair ; the former with the clypeus acute and prominent at the angles, and reflexed along the entire margin ; the latter with the median line distinct almost throughout. Elytra of a rather dark red, subnitid, sparingly pilose and coarsely and irregularly punctate with the geminate striæ distinct. Pygidium, under surface of body, and legs, red and cinereo-pilose.

378.—LIPARETRUS TRIDENTATUS. n. sp.

Length 2½ lines.

Head black, punctate, and clothed with long yellow hair, with the clypeus reflexed, and armed in front with three teeth. Thorax black, opaque, subsericeous, finely punctate and moderately flavo-pilose. Elytra of an opaque subsericeous red, with the punctures rather shallow and the geminate striæ tolerably distinct. Pygidium and under side of body black and moderately pilose. Legs piceous.

The form of the clypeus in this species is remarkable, but in other respects I see nothing which should remove it from that section of *Liparetrus* which has the antennæ eight-jointed.

379.—LIPARETRUS GLABER. n. sp.

Length 2¼ lines.

Head and thorax black, subnitid, punctate, and free from hair, the former with the clypeus reflexed and somewhat rounded in front. Elytra reddish yellow, nitid, free from hair and irregularly punctate with the geminate striæ traceable. Abdomen reddish

M

yellow, finely punctate and very sparingly pilose. Pygidium large. Legs piceous brown, with the external angle of the fore tibiæ prolonged to half the length of the tarsi, and without external teeth.

380.—LIPARETRUS PARVULUS. n. sp.

Length 1¾ lines.

This insect only differs from the last in being smaller, in having the pygidium and under side of body black, and in having a narrow lateral black margin on the elytra. Both species have the same peculiarly formed fore tibiæ, more like those of the genus *Diphucephala* than of *Liparetrus*.

381.—SCITALA SUTURALIS. n. sp.

Length 6 lines.

Head and thorax black, finely punctate, the latter pruinose in the centre, and reddish on the sides. Elytra pale testaceous with the suture black, subnitid, slightly silky and punctate, with a sutural and lateral stria, and four geminate striæ on each elytron. Under side of body and legs piceous red, nitid and finely punctate.

This species is not unlike the *S. pruinosa* of Dalman. The puncturation throughout however is much finer, and on the thorax less dense, and the elytra though slightly silky are without the opaque velvety texture of *pruinosa*.

382.—SCITALA ARMATICEPS. n. sp.

Length 5¼ lines.

Head and thorax piceous brown, finely and densely punctate ; the former with a strong transverse ridge on the clypeus ; the latter short, and reddish at the sides. Elytra testaceous with the suture brown, nitid, and punctate with a sutural and lateral stria and four geminate striæ on each elytron. Pygidium, under side of body and legs, pale red, nitid and finely punctate.

HOMOLOTROPUS. n. gen.

Antennæ of nine joints : 1st, long, clavate ; 2nd. globular ; 3rd. slight and cylindrical ; 4th. short ; the remaining five joints forming a club with the laminæ long. Labrum transverse, deeply emarginate. Mentum rough, broad at the apex and slightly emarginate. Palpi with the terminal joints a little longer than the others. Suture of clypeus indistinct. Thorax lobed behind. Elytra not covering the pygidium. Fore tibiæ strongly tridentate. Body oblong flat.

The position of this genus is evidently near the genus *Xylonychus* MacLeay. That genus seems to me to be more naturally placed among the *Heteronycidæ* of Lacordaire, than as he has placed it, among the true *Melolonthidæ*.

383.—HOMOLOTROPUS LURIDIPENNIS. n. sp.

Length 8 lines.

Head dark brown, with the clypeus large, triangularly extended behind, and very densely punctate, and with the forehead subnitid and very sparingly punctate. Thorax of a testaceous red clouded with black, thinly punctate in patches, transverse, moderately emarginate in front, rounded on the sides, and roundly lobed at the base, with a carination on the basal half of the median line. Scutellum brown, large, densely punctate, rounded at the sides,and pointed at the apex. Elytra of a pale brownish yellow, subnitid, broader than the thorax, almost truncate at the apex, sparingly pilose and punctate, with a sutural striæ and four geminate striæ on each elytron, and irregular punctures and small patches of a dark brown colour as in the elytra of *Anoplognathus inustus* of Kirby. Body piceous brown, with the pygidium and sides of the abdominal segments densely clothed with a cinerous pubescence. The under side of the thorax and thighs are pilose.

384.—HAPLONYCHA PINGUIS. n. sp.

Length 8 lines.

Reddish chesnut, nitid, very finely punctate, of oval form and very thick. Head almost brown in front and rugosely punctate.

Thorax without trace of median line, and with a shallow fovea near the sides at the broadest part. Elytra with the sutural and four geminate striæ distinct. Pygidium large, completely uncovered, smooth, nitid, and mottled black and red. Abdomen piceous brown, smooth, nitid. Pro-meso and metathorax flavopilose.

This species differs from *H. obesus* in being much more finely punctate, and in having the pygidium much larger, smoother and completely exposed.

385.—HETERONYX HOLOSERICEUS. n. sp.

Length 6 lines.

Pale chesnut, subopaque, very finely and densely punctate, and clothed very thickly with a silky pale yellow pubescence. Head a little darker and more coarsely punctate than the rest of the upper surface, with the clypeus entire, and the labrum not appearing in front of it. Thorax ciliated at the sides with soft reddish hairs, and almost truncate at the base, with an indistinct blackish spot accompanied by a few coarse punctures near each side. Elytra striate,—the sutural stria distinct, the rest almost obsolete—and ciliated at the sides with soft reddish hairs. Pygidium exposed, roughly punctate, and sparingly pilose, with a fovea on each side and an indistinct depressed line in the middle. Under surface less densely pubescent than the upper. Legs red.

386.—HETERONYX PUBESCENS. n. sp.

Length 5½ lines.

Brownish red, subnitid, punctate and pubescent. Clypeus emarginate with the apex of the labrum visible from above. Thorax and elytra coarsely and almost rugosely punctate with the pubescence of a yellow colour and longer than in *H. holosericeus*. Body beneath and pygidium very sparingly pubescent.

387.—HETERONYX CASTANEUS. n. sp.

Length 4 lines.

Pale chesnut, subnitid, finely punctate and pubescent. Clypeus emarginate with the narrow and rounded apex of the labrum visible in front of it. Thorax and elytra much more finely punctate than the last species, the latter with a few almost obsolete traces of striæ. Pygidium and body beneath piceous and moderately pilose.

388.—HETERONYX SUBSTRIATUS. n. sp.

Length 3½ lines.

Brown, subnitid, punctate and pubescent. Clypeus slightly emarginate, with the apex of the labrum broad and truncate, and visible in front of it. Thorax not densely punctate, and with the median line distinctly impressed. Elytra densely punctate and moderately pubescent, with the striæ on the elytra more distinct than in any of the preceding species. Body beneath rugosely punctate.

389.—HETERONYX INFUSCATUS. n. sp.

Length 3½ lines.

Brown, nitid, punctate—the punctures large and thinly placed —and pilose. Clypeus densely and rugosely punctate, and truncate in the middle, with the apex of the labrum broad, very slightly rounded and visible in front of it. The sutural stria on the elytra is pretty distinct. The abdominal segments are coarsely punctate and finely acuducted. The legs are red.

390.—HETERONYX PALLIDULUS. n. sp.

Length 3½ lines.

Elongate, pale red, subnitid, punctate, without pubescence. Clypeus rugosely punctate, and truncate in front, with the apex of the labrum broad, nearly truncate, and visible in front of it. Thorax somewhat convex, and thinly and finely punctate, with a small brown tuberosity near each side, and the median line traceable only in the centre. Elytra covered with irregular shallow punctures, and with the sutural and one or two other striæ tolerably defined. Body beneath very sparingly pilose.

391.—HETERONYX CONCOLOR. n. sp.

Length 2¾ lines.

Flat, brownish red, subnitid, punctate and finely pubescent. Clypeus slightly emarginate with the apex of the labrum broad and appearing in front. Thorax finely, elytra roughly punctate. Body beneath very sparingly pilose.

392.—HETERONYX RUFICOLLIS. n. sp.

Length 2½ lines.

Elongate, punctate, pubescent, subopaque, and black, with the head thorax and legs red. Clypeus slightly emarginate with the apex of the labrum broad and visible in front. Thorax rounded at the base and posterior angles. Elytra subtruncate and obsoletely striated.

393.—HETERONYX RUGOSIPENNIS. n. sp.

Length 6 lines.

Reddish chesnut, nitid, thinly punctate, and clothed with erect fulvous pile. Clypeus rounded, reflexed and entire. Thorax almost truncate at the base, and with a brown nodule near each side. Scutellum densely punctate except at the extremity. Elytra coarsely punctate and transversely rugose. Pygidium and under side of body thinly punctate—the punctures large and each furnished with a long erect hair.

I have some doubts about this being a *Heteronyx* at all. The labrum is very short, almost truncate, and scarcely turned up at the apex.

ODONTOTONYX. n. gen.

Antennæ of nine joints, 1st long, clavate ; 2nd globular ; 3rd, 4th, 5th, and 6th very small ; the other three forming a club with the laminæ short. Labrum short, narrow, horizontal, lightly emarginate, and completely covered by the clypeus. Mentum apparently rounded at the apex. Palpi with the terminal joint longest. Elytra not entirely covering the pygidium. Fore tibiæ tridentale. Ungues of tarsi strongly toothed beneath. Body oblong, flat.

394.—ODONTOTONYX BRUNNEIPENNIS. n. sp.

Length 4 lines.

Head red and densely punctate, with the clypeus round and reflexed in front. Thorax of a paler red, thinly and finely punctate, and rounded a little at the base, with the posterior angles obtuse, and a small brown fovea near each side. Scutellum red, oblong and rounded at the apex. Elytra brown, nitid, profoundly striated, and roughly punctate. Pygidium pale red, opaque, very finely punctate and moderately pilose. Under side of body and legs of a pale testaceous colour, nitid, thinly punctate and moderately pilose, and on the legs setose. Teeth of fore tibiæ strong and of a brown colour.

In the only specimen of this insect in the collection there seems to be a kind of membranaceous appendage beneath to the last joint of the tarsi.

Sub-family. RUTELIDÆ.

395.—ANOPLOGNATHUS LINEATUS, MacL., W. *Trans. Ent. Soc. N. S. Wales*, 1864, *page* 18.

396.—REPSIMUS PURPUREIPES. n. sp.

Length 10 lines.

This may be only a variety of *R. æneus.* It is, however, a considerably larger insect, has the pygidium more exposed, the scutellum more distinctly punctate, the elytra of a purplish black, and the legs of a splendid reddish purple.

Sub-family. DYNASTIDÆ.

397.—ISODON PUNCTICOLLIS. n. sp.

Length 7½ lines.

Reddish brown, nitid. Head black, densely punctate, and rugose, with a transverse ridge at the suture of the clypeus, with the clypeus subtriangular, emarginate at the sides with a raised margin and two small tubercles in front, and with the mandibles externally tridentate. . Thorax convex, subtransverse, rounded on the sides and posterior angles, almost truncate at the base, and thinly punctate except near the anterior margin, with a small tubercle, and a round excavation behind it on the anterior margin in the male, and an almost obsolete impression in the same place in the female. Scutellum large, smooth and rounded behind. Elytra with the suture brown, with a deep stria on each side of it, and with the rest of their surface covered with irregular rows of coarse punctures, the punctures becoming smaller and denser towards the sides and apex. Legs and under side of body piceous, nitid and fulvo-pilose.

398.—ISODON LÆVICOLLIS. n. sp.

Length 8 lines.

Head and clypeus brown, transversely rugose, the latter with raised margins and two small tubercles in front as in the last species, and with the mandibles externally tridentate, the posterior tooth much larger than the others. Thorax convex, piceous red, nitid and smooth. Scutellum also piceous red, large, rounded and smooth. Elytra black and nitid, with the suture piceous red, with a stria on each side of the suture, and with the rest of their surface covered with rows of large punctures much as in the last species but more regularly disposed. Pygidium smooth. Legs and under side of body piceous red, and moderately pilose.

The only specimen I have seen is a female.

399.—HETERONYCHUS PICIPES. n. sp.

Length 8½ lines.

Black, nitid. Head rugose with two small tubercles on the transverse ridge, with the clypeus subtriangular and emarginate in front, and with the mandibles toothed as in the last species. Thorax convex, smooth, and very minutely punctate, with a very slight depression in front in both sexes. Scutellum smooth, rounded at the apex. Elytra with a deep stria on each side of the suture, a smooth space beyond these striæ, punctured striæ over the rest of the surface, and a dense mass of small punctures at the apex. Pygidium punctate, smooth in the middle. Legs and under side of body piceous and pilose.

400.—HETERONYCHUS IRREGULARIS. n. sp.

Length 8 lines.

Black, subnitid. Head very rugose, with two indistinct tubercles on the transverse ridge, with the clypeus subtriangular and minutely trituberculate in front, and with the mandibles toothed as in the last two species. Thorax not punctate, with a tooth on the anterior margin and a round excavation behind it in the male,—the female I have not seen. Scutellum as in the last species. Elytra with a deep stria on each side of the suture, the space beyond, which in the last species was smooth, marked with large irregularly placed punctures, and the rest of the surface covered with irregular rows of large punctures. Pygidium smooth. Legs and under side of body piceous red and sparingly pilose.

401.—DASYGNATHUS MASTERSII. n. sp.

Length 10 lines.

Black, nitid. Head densely punctate, and furnished in the male with a blunt slightly recurved short horn. Thorax very finely and thinly punctate, without any impression in the female, but in the male retuse and deeply excavated in front, with two rather acute tubercles at the summit of the centre, an obtuse protuberance at the lateral boundary of the retuse portion, and outside of these protuberances an elongate punctate fovea. Scutellum broadly rounded, punctate at the base. Elytra piceous brown, nitid, and covered with large shallow striæ, filled with large shallow punctures, the striæ becoming less marked and the punctures smaller and more numerous towards the sides and apex. Pygidium very minutely punctate. Under surface of body piceous. Sides of head, thorax, elytra above and entire under side of body and legs ciliated with red hair.

402.—ORYCTES OBSCURUS. n. sp.

Length 10 lines.

Of a dull chocolate colour, very opaque. Head densely punctate with a somewhat acute slightly recurved horn in the centre of the forehead, and with the clypeus subsinuate at the sides and narrow, truncate and bituberculate in front. Thorax largely retuse in front, marked on the retuse portion and along the anterior parts generally with large shallow transverse variolous looking impressions, and on the basal portion with small distant punctures. Scutellum rugosely punctate, smooth on the margins and semicarinate in the middle. Elytra little convex, and apparently smooth but seen under a lens to be marked with light striæ, each stria composed of very shallow variolous impressions. Pygidium coarsely punctate. Under side of body fulvo-pilose.

The only specimen of this insect is a male.

403.—SEMANOPTERUS DEPRESSIUSCULUS. n. sp.

Length 9 lines.

Black, subnitid. Head rugosely punctate with a small conical tubercle in the centre of the forehead, and the apex of the clypeus slightly rounded in front, and reflexed in the middle into a minute tooth. Thorax transverse, deeply emarginate in front, rounded on the sides, almost truncate at the base, with the anterior angles advanced and acute, the posterior square, and the median line marked with two punctate foveæ, the anterior one round, the posterior elongate. Scutellum punctate at the base, smooth at the apex. Elytra with the suture smooth and elevated, and with four costæ on each side, the intervals between the first three having three irregular rows of coarse punctures, while towards the sides the puncturation becomes closer and much less regular. Pygidium finely punctate. Under surface nitid, with the meso-and meta-thorax densely punctate, and fulvo-pilose. Fore tibiæ strongly tridentated externally and acutely so in the males, and with the internal apical spine large in both sexes and acute in the male.

404.—SEMANOPTERUS CONVEXIUSCULUS. n. sp.

Length 7½ lines.

Black, subnitid, subconvex. Head transversely rugose, with a small tubercle in the centre of the forehead, and with the clypeus truncate. Thorax finely punctate with the impressions on the median line more shallow, more elongate and less punctate than in the last species. Scutellum subtriangular and smooth with two or three punctiform impressions in the centre. Elytra strongly striato-punctate, with the interstices rather elevated, and the punctures towards the apex small and dense. Under surface and legs as in the last species.

405.—CRYPTODUS SUBCOSTATUS. n. sp.

Length 9 lines.

Brownish black, subnitid, and of a broad flattish form. Head densely punctate, with two small tubercles on the forehead, and with the clypeus rounded and reflexed. Thorax subconvex, finely but not densely punctate, with the median line broadly but not deeply impressed. Elytra flat, of the width of the thorax, and slightly dehiscent, with the suture and three tolerably distinct lines on each elytron elevated and thinly punctate, with the intervals occupied by three irregular rows of oval variolous impressions, and with a strong tuberosity near the apex on each elytron. Under side of body and legs piceous, thinly punctate, each puncture bearing a very short yellow decumbent seta. Antennæ with the first joint rough, punctate, triangular and very large, completely covering the rest of the antennæ excepting the club. Mentum with the base deeply and rather narrowly emarginate.

406.—CRYPTODUS OBSCURUS. n. sp.

Length 8½ lines.

Differs only from the last in being opaque, in having the clypeus less reflexed, the thorax more punctate and less convex, the elytra shorter, and the form generally much narrower and rather more convex.

407.—CRYPTODUS INCORNUTUS. n. sp.

Length 8 lines.

Piceous brown, subnitid, subconvex, narrow. Head densely punctate, with the clypeus broad, slightly reflexed and almost truncate. Thorax coarsely and densely punctate, with the median line broadly marked in the middle. Elytra dehiscent, with the subapical tuberosities strong, the costæ scarcely elevated, and the variolous impressions in the intervals of an elongate oval shape. Under side of abdomen densely variolose-punctate. First joint of antennæ a little smaller than in the last two species. Base of mentum deeply emarginate. Apex of prosternum subtruncate and ciliated.

These three species differ from all those previously described in the great size of the first joint of the antennæ. I may here state that *C. paradoxus* MacLeay, of which I possess the original specimen is a very different insect from *C. variolosus* White.

Sub-family CETONIIDÆ.

408.—SCHIZORHINA IMPAR, MacL., W. *Trans. Ent. Soc. N. S. Wales*, I., 1863, *page* 14.

409.—SCHIZORHINA OCELLATA, MacL., W. *Trans. Ent. Soc. N. S. Wales*, I., 1863, *page* 16.

410.—SCHIZORHINA PUNCTATA, Don. *Epit. Ins. t.* I— Gory et Percheron. *Mon. page* 164, *t.* 28. *f.* 4,—Burm. *Handb.* III., *page* 541.

411.—SCHIZORHINA AUSTRALASIÆ, Don. *Epit. Ins. t.* I. *f.* I.—Gory et Percher. *Mon. page* 161. *t.* 28. *f.* I. Burm. *Handb.* III., *page* 650. *Orpheus* Perron *i. litt.* *Pauzeri* Swartz. *Schönh. Syn. Ins.* I., 3, *app., page* 50.

412.—SCHIZORHINA MASTERSII. n. sp.

Length 8½ lines.

Black, nitid. Head and thorax densely punctate, the former

with a transverse elevation highest in the middle on the forehead, and with the clypeus lightly emarginate. Scutellum thinly punctate. Elytra piceous towards the sides, and roughly and irregularly striato-punctate, with the second interstice from the suture, and the humeral callosity enlarged. Abdomen beneath very finely and thinly punctate with the apex of the penultimate segment ciliated with reddish hair. Apex of the mesosternum keel shaped, not produced. Under side of thorax and thighs moderately hairy. Tibiæ and tarsi red.

413.—Schizorhina hirticeps. n. sp.

Length 7 lines.

Head and thorax black, densely punctate and pilose, the former with a ridge on the forehead, the base of the clypeus yellow, and the apex slightly emarginate, the latter with the apex, sides excepting a small spot, and two elongate spots near the base, yellow. Scutellum thinly punctate, and brown with the centre yellow. Elytra glabrous with the suture elevated and of a dark brown colour, with a broad elevated space on each side of the suture marked with one row of punctures and of a yellow colour with a brown spot behind the middle, and with the rest of their surface rugosely punctate, and of a luteous colour. Pygidium yellow, with the margins and a broad median vitta, brown. Abdomen beneath very finely and thinly punctate, nitid, and of a yellow colour with a black spot on the side and in the centre of each segment. Pro-meso-and meta-thorax black spotted with yellow, densely punctate and clothed thickly with long greyish hair. Apex of mesosternum, rounded and yellow. Legs red.

414.—Schizorhina nigrans. n. sp.

Length 7 lines.

In shape, size, and sculpture, this species is identical with the last, but it is entirely of a nitid black, with the head, thorax, and under side of body, thickly clothed with greyish hair.

415.—Schizorhina pulchra. n. sp.

Length 7 lines.

Olive brown with a tinge of red, opaque, velvety. Head

glabrous, subnitid, and punctate with the clypeus strongly emarginate. Thorax thinly punctate the punctures of a semicircular form, with the sides broadly margined with yellow. Scutellum long, obtuse at the apex. Elytra thinly punctate as on the thorax, with some fine striæ near the suture, numerous transverse striæ near the apex, and an interrupted yellow fascia behind the middle. Pygidium black with two large yellow spots. Under surface of body and legs piceous, punctate, each puncture furnished with a white setiform scale, and marked with numerous whitish patches. Thighs and tibiæ ciliated beneath with whitish hair. Apex of mesosternum broad and rounded.

416.—Schizorhina viridicuprea. n. sp.

Length 6 lines.

Of a brilliant coppery green, with the upper surface of the thighs, the tibiæ, the tarsi, the antennæ, and the parts of the mouth. of a piceous red. Head long, finely punctate, with the clypeus slightly emarginate, and of a bluish colour at the apex. Thorax smooth and rather elevated along the median line, depressed on each side of it, and marked with a number of short transverse striolæ. Scutellum with a few punctures near the sides. Elytra furnished with numerous erect hairs, punctured in irregular rows near the suture, rugosely striolate towards the sides, and impressed with a large oblong fovea behind the middle. Under side of thorax and thighs pilose. Apex of mesosternum flat, dilated and rounded.

The sculpture of this insect is the same as in *Cetonia fulgens* described by me in Vol. I., page 18, of our Transactions, but it differs from it in being of a broader form, and very different colour.

417.—Glycyphana brunnipes, Kirby. *Trans. Linn. Soc.* XII., *page* 465,—Schaum. *Ann. Franc.* 1849, *page* 263.
conspersa Gory et Perch. *Mon.*, *page* 287, *t.* 56, *v.* 1.—Burm. *Handb.* III., *page* 353.
obscura Don. *Epit. Ins ?*
viridiobscura, Dej. *Cat.* 2nd *ed.*, *page* 173.
var *fasciata.* Fab. *perversa*, Schaum. *stolata.* Fab.

418.—VALGUS NIGRINUS. n. sp.

Length 1½ lines.

Entirely black, and moderately covered with white scales. Head punctate with the clypeus broadly rounded at the apex. Thorax longer than the width, and covered with a kind of tesselated marking. Scutellum thickly covered with white scales. Elytra punctate, striate, and marked much in the same way as the thorax.

419.—VALGUS CASTANEIPENNIS. n. sp.

Length 1½ lines.

Head and thorax black, densely covered with whitish scales, the former with the clypeus truncate in front. Elytra chesnut and covered with striæ of yellowish scales. Pygidium under side of body, and legs very densely clothed with whitish scales.

The remainder of the Coleoptera in the Gayndah collection will form the subject of another paper.

On Australian Entozoa, with descriptions of new species, by

GERARD KREFFT, F.L.S.

[Read July 3rd, 1871.]

THE natural history of the intestinal worms has been much neglected in Australia, and we are not yet able to tell how many of the common European species have accompanied man and his domestic animals to this country. It has been ascertained on the other hand that some Entozoa, tape-worms for instance, of purely Australian origin infest our sheep, and it is a well known fact that the common sheep fluke (*Fasciola* or *Distoma hepatica*) has long occupied the biliary ducts of the kangaroos. Our rats are troubled with cysts which contain unmistakeable young tape-worms, (*Cysticercus fasciolaris*), and nearly every specimen which I examined carried several of them, from a few lines to two inches in length. It would be interesting to know in which animal this cestoid attains maturity.

Our water fowl are great cestoid bearers, and as these birds are easily obtained, I have first paid attention to these and now lay the result of my investigations before the Entomological Society of New South Wales.

New genera were not discovered; some of our species I found to be closely allied to European ones, (such as the Hammer-headed tape-worm (*Tænia malleus*) ; the young of other species were traced, and it was observed that they lived in prodiguous numbers in the hosts which carried the perfectly mature " colony." My observations were made at first without the proper means, I had no microscope, and was without a standard work on the subject, both deficiencies were, however, kindly supplied by gentlemen interested in these researches, so that the next shooting season will find me better prepared to make correct drawings of the ova, and take the necessary measurements.

The ova have assisted me much in arriving at a final conclusion as to the character of the variable collection obtained.

I had jotted down and sketched, what I considered at least thirty species, and these in a great measure owing to the ovutest, were reduced to fifteen or sixteen. My practical acquaintance with Entozoa is of a recent date, but I find the subject of such interest that every moment which could be spared from other duties has been devoted to it. There is no necessity to point out the importance of the study of this group of animals to the well being of millions, but as many people consider it a particularly nasty subject, I will try and prove to them that it is not so.

The fresh intestines are put into a flat dish, and a stream of water is kept running over them till quite clean, a rough board beneath prevents the escape of any of the smaller Entozoa which the water may force out. The parts are then opened and after a gentle flow of the element for several hours, the worms may be picked out with a camel hair brush.

Not having time to make the usual preparations for the microscope, and being generally well supplied with duplicate specimens, I spread some out on glass slides, and observed them, making the necessary sketches at the same time ; they dried gradually and the changes which the different parts underwent were carefully noticed. Whenever I wish to refer to a specimen it is put under the microscope, and any part, even if it is a cestoid of several feet in length, can be examined without difficulty. To clear up doubtful points, some wet preparations are necessary, and these are kept in glass tubes to be used when the first plan fails.

When once dry, the objects are transparent, and under the glass look most charming—the fact is, few persons unacquainted with them will believe what they are. Some retain their colour, and are therefore still more valuable.

I have mounted the largest flukes in the same way, but had to keep them for a day or so in water and press them slightly when they were too thick. With a view of giving some idea of the arrangement of the Entozoa, I have added Professor von Siebold's system, and I need not observe that his first order of *Cystici* is now generally accepted as part of the Cestoidea ; the genera arranged under that head being no doubt young tapeworms.

N

Class. HELMINTHES.

It is very difficult to characterize the class Helminthes, for it contains animals having widely dissimilar organization. On this account the separation of its groups and their distribution among the other classes of the invertebrata has been attempted. But such various difficulties have arisen from this, that for the present it is best that all these animals should remain together. If a common character is not furnished by their structure, it must be sought for in their manner of life; for nearly all are parasites, and during their whole life or at least during some of its periods, seek their abode and nourishment in or upon other living animals.

Order I. Cystici.*

The body is swollen in form of a bladder, and filled with a serous liquid. Digestive and genital organs are wanting.

Genera : *Echinococcus, Cænurus, Cysticercus, Anthrocephalus.*

Order II. Cestodes.

The parenchymatous body is riband like, having often incomplete transverse fissurations; often it is wholly divided transversely into rings. Digestive organs are wanting. The genital organs of both sexes are combined in the same individual, and generally are often repeated. Copulatory organs are present.†

Genera : *Gymnorhynchus, Tetrarhynchus Bothriocephalus, Tænia, Tricænophorus, Ligula, Caryophyllæus.*

---

* Nearly all the genera of their order are considered to be young tape-worms.—G.K.

† Von Siebold has changed his opinion, and states some years later "On the Tape and Cystic worms." (translated by Professor Huxley, London, 1857, page 40)." The sexually matured individuals of the *Cestoidea* are no other than their full grown joints in which are developed the male and female genitalia, by whose co-operation eggs capable of reproduction are generated, and the continuation of the species is secured. Such a sexually-mature hermaphrodite joint of a cestoid worm which separates from the body of the scolex with great readiness, is denominated a *Proglottis*. The formation

Order III. Trematodes.

The body is parenchymatous and usually flattened. The intestinal canal, which is often branching, has a mouth, but nearly always is without an anus. The genital organs of both sexes are combined in the same individual. Copulatory organs are present.

Genera : *Gyrodactylus, Axine, Octobrothrium, Diplozoon, Polystomum, Aspidocotylus, Aspidogaster, Tristomum, Monostomum, Holostomum, Gasterostomum, Pentastomum.*

Order IV. Acanthocephali.

The sack-like body is flattened, transversely striated, and swollen cylindrically by the absorption of water. Digestive organs are wanting. The genital organs are situated in separate individuals. Copulatory organs are present.

Genus : *Echinorhynchus.*

Order V. Gordiacei.

The body is filiform and cylindrical. The digestive organs

---

of these *Proglottides* takes place at the posterior end of the scolex by a sexual reproduction, viz., by a simple process of growth and division. If we compare this process with the phenomena of the alternation of generations, we shall discover in it all the essential characters of the latter. The matured joints or the sexual individuals of the *Cestoidea* in their proglottis form, produce a brood of embryos armed with six hooklets (see plate III., figs. 22 and 23 of this paper—G.K.) which are quite dissimilar in shape from their parents, the Proglottides, (see plate III., fig. 21b of this paper—G.K.) and remain so, since at a later period they assume the scolex form, and take on the functions of an agamozooid. From the posterior end of the body of such a scoliciform agamozooid a series of joints are developed,—that is to say, a generation of sexual individuals which again present the original proglottis form. In their organization the *Proglottides*, apart from their sexual apparatus, so far resemble the scolices from which they have been produced, that they possess no oral aperture, and moreover are subject to a deposit under their integument of those glassy calcareous particles which I have already mentioned. (A good idea of a young scolex will be obtained by comparing the 2nd plate, figs. 2b and 2c, with the adult form figs. 2 and 2a—G.K.) It seems, at first, paradoxical to say that the joints of a tapeworm which have hitherto been believed to be mere parts of one animal, should be considered as individuals ; but whoever will observe with are without an anus. Genital organs are situated upon separate individuals. Copulatory organs are sometimes present. Genera : *Mermis, Gordius.*

Order VI. Nematodes.

The body is sack-like and cylindrical. The digestive canal has a mouth and anus, and passes in a straight line through the cavity of the body. The genital organs are situated upon separate individuals. Copulatory organs are present.

Genera : *Sphaerularia, Trichosoma, Trichocephalus, Filaria, Anguillula, Physaloptera, Liorhynchus, Lecanocephalus, Cheiracanthus, Gnathosoma, Ancyracanthus, Spiroptera, Hedruris, Strongylus, Cucullanus, Oxyuris, Ascaris.*

I consider it also necessary (for the purpose of making those interested better acquainted with what has been written on Australian Entozoa) to give a complete list of the species already described. This I propose to do in chronological order.

The first Australian intestinal worm (*Tænia festiva*) was noticed by Rudolphi in the year 1819, in his " Entozoorum

an unprejudiced eye, a fully developed *Tænia* with its sexually matured joints, must be convinced that it is no simple animal, but one composed of many individuals."

On page 44: von Siebold remarks :—" In the *Cestoidea* the stock is the posterior end of the scoleciform agamozooid (the head G. K.) In the alternation of generations amongst the *Cestoidea*, there is this peculiarity that the agamozooid preserves its efficacy and independence, whilst the agamozooids of other animals which undergo alternation either die after producing their brood or pass into it. (Huxley doubts this : G.K.) We must consider the head of every cestoid worm as the agamozooid still remaining and capable of reproduction, and its neck as the equivalent of the posterior extremity of the scolex. In all cestoids we see that fresh joints are continually being developed at the posterior part of the neck which lengthens and becomes covered with transverse folds. These folds are at first very close together, but as the process of growth throws them backwards further and further from their place of origin, they gradually change from indistinct wrinkles into sharp transverse lines of demarcation, between which the substance of the body dilates into a joint (individual), and assumes its specific shape. At a later period the rudiments of the hermaphroditic sexual apparatus make their appearance in the interior of the joints, and finally they separate themselves from their younger fellows as independent individuals."

Synopsis," page 146, from the intestines of the Great Kangaroo (*Halmaturus giganteus*.) The second discovery was, that the common sheep fluke (*Distoma* or *Fasciola hepatica*) inhabited the biliary ducts, and the liver of Kangaroos (Ib. page 725). No description of Australian species occurs till February 8th, 1853. (Proceedings of the Zoological Society of London for 1853, pages 18 to 25), when Dr. Baird describes the following species :—

Ascaris similis.—From the stomach of a Seal.

Mermis rigidus.—Habitat unknown but probably Australian.

Tœnia Goezii.—Habitat unknown, probably Australian.

Tœnia Cederi.—Habitat. The stomach of a Penguin from the Antarctic seas.

Bothriocephalus antarcticus.—Habitat. The stomach and intestines of a Southern seal.

In the year 1859, Dr. Baird described (Proceedings of the Zool. Soc. of London, page 111). *Tœnia sulciceps*, from the intestines of the Wandering Albatross (*Diomedea exulans*), and also noticed a rare species of *Ascaris* from the Dugong described by Professor Owen as *Ascaris halichoris*, (Ib. pages 148 and 149).

In 1861, Dr. Baird noticed a small Filaria (*Filaria sanguinea*) in the stomach of a little Australian fresh water fish (*Galaxias scriba*).

In the same publication for 1862, I find a description of a new Pentastoma (*Pentastoma teretiusculum*) by Dr. Baird, (page 114) who mentions that he took the specimen from the mouth of a snake (*Hoplocephalus superbus*), which died at the Zoological Society's gardens. I may state here that this worm is generally found in the lung of Australian snakes.

The last of Dr. Bairds descriptions occurs in the Proceedings of the Zoological Society for 1865, page 58, and relates to a new cestoid worm (*Bothridium* (*Solenophorus*) *arcuatum*.) This species is common in the Australian Diamond snake (*Morelia spilotes*).

In that most useful book the "Zoological Record," I find mention made of an Australian tape-worm from the stomach of the Emeu (*Dromaius novæ-hollandiæ*), which is described by the Danish Naturalist Krabbe (Record for 1869, page 635), as *Tœnia australis*.

The following species were taken by myself and by Mr.
George Masters in the neighbourhood of Sydney and in Queensland :—

MAMMALIA.

From a Dolphin (*Delphinus Forsteri*) —
1. A species of Tænia.
2. A species of Distoma.
3. An Ascaris.
4. An Echinorhynchus.

From a Kangaroo (*Macropus major*) —
1. A Fluke (*Fasciola hepatica*).

From a Bandicoot (*Perameles nasuta*)—
A species of *Ascaris*.

From a domestic Cat (*Felis catus*)—
An Ascaris (*Ascaris mystax*).

From a Sheep (*Ovis aries*)—
1. *Amphistoma conicum* (?)
2. *Strongylus filaria*.
3. *Ascaris* spec. (?)

From a Pig (*Sus scropha*) —
1. *Strongylus suis*.

From a Wallaby (*Halmaturus*)—
1. A species of *Tænia*.
2. A second species of *Tænia*.

From a Phalanger (*Phalangista vulpina*)—
1. A species of *Tænia*.

AVES.

From a black Duck (*Anas superciliosa*)—
Several species of cestoid worms.

From a Bower Bird (*Chlamydera maculata*)—
1. A species of *Tænia*.

From a Pigeon (*Columba livia*)—
1. A species of *Ascaris*.

From a White-eyed Duck (*Nyroca australis*)—
1. A cestoid worm, a quarter of an inch and more in width, and several feet in length.
2. Several other smaller species of *Tænia.*

From a Teal (*Anas punctata*)—
1. Several cestoid worms, including the young of the Australian Hammer-headed Tape-worm.

From a Porphyrio or Red-bill (*Porphyrio melanotus*)—
1. A very small species of *Distoma.*

From a White Crane (*Herodias alba*)—
1. A very large Distoma which is distinguished from all others in having the vaginal papilla below the ventral pore, and not as usual between this pore and the oral sucker.

From a Pacific Crane (*Ardea pacifica*)—
1. A species of *Distoma.*

From a Snake Bird or Darter (*Plotus novæ-hollandiæ*)—
1. A new species of *Ascaris.*

From a Little Grebe (*Podiceps australis*) —
Several species of *Tænia.*

From a Stilt (*Himantopus leucocephalus*)—
Two cestoid worms.

From a Shoveller Duck (*Spatula rhynchotis*)—
A very curious short cestoid worm with very large head.

From a Musk Duck (*Biziura lobata*)—
A fine species of cestoid worm.

From a Gill Bird (*Anthochæra carunculata*)—
A series of *Ascari* taken from the eye by Mr. George Masters.

REPTILIA.

From a Tortoise (*Elseya dentata*) inhabiting northern Rivers—
1. A species of *Amphistoma* taken by Mr. George Masters.
2. A species of *Ascaris.*

From a Lizard (*Egernia Cuninghami*)—

    1. Numerous very small and still undetermined Round Worms, some with long pointed tail.

From a Cyclodus or Sleeping Lizard *(Cyclodus gigas)* —

    1. A species of *Physaloptera*.

From a White's Hinulia (*Hinulia Whitei*)—

    1. A species of *Ascaris*.

From a Ribbon Lizard (*Hinulia tœniolata*)—

    1. A species of *Pentastoma*.

From a Gecko (*Diplodactylus ornatus*)—

    1. A species of *Pentastoma*.

From a Gecko (*Phyllurus Miliusii*)—

    1. A species of *Ascaris*.

From a Diamond Snake (*Morelia spilotes*)—

    1. A species of *Bothridium*, probably identical with *Bothridium arcuatum* Baird.

    2. A species of *Ascaris* with very long spiculæ.

From a Carpet Snake (*Morelia variegata*)—

    1. A species of *Ascaris* identical with the above.

From a Grey Snake (*Diemenia reticulata*)—

    1. Males and females of a large *Pentastoma*, found attached to the lungs.

    2. A species of *Physaloptera*.

    3. An *Echinorhynchus*.

PISCES.

From a large vegetable feeding Percoid Fish

    1. A species of *Distoma*.

    2. A Round Worm probably a species of *Trichocephalus*.

From a "Forster's Ceratodus" (*Ceratodus Forsteri*)—

    1. A species of *Ascaris*, obtained by Mr. George Masters.

The following descriptions are those of the new species :—

TÆNIA TUBERCULATA.

(Plate I., figs. 9, 10, 11, 12, 13, 14, 15, 16, 17, 18, 19, 20, 21, 21a.
Plate II., fig. 3).

Total length seldom exceeding 42 inches, average breadth
one quarter of an inch, specimens occur however, which are half
an inch wide in some parts. Head small, variable, often attached
to a long slender neck which more frequently ends in a filament.
Tapering specimens such as figured on Plate II. (fig. 3), are
common ; the neck appears quite perfect but not a vestige of a
disk, sucker or proboscis can be observed. I examined more than
25 White-eyed Ducks (in which this tape worm is principally
found), and obtained over fifty specimens but only five or six
were furnished with heads such as are sketched (much enlarged)
on Plate I. (figs. 12, 13, 14, 16, 19, 20, 21, and 21a). The
general form of this cestoid resembles fig. 9 of Plate I. Owing to
the many tubercles distributed over the posterior portion of the
segments, the appearance of the colony is irregular though the
marginal lines are generally straight. The anterior portion for
about one-fifth of the total length is provided with very close
segments directly after which the lemniscy appear, one on each
side of every joint, the edges enlarging till they look in the most
posterior proglottides, like small mammæ.

Plate II., fig. 3, shows the size of the segments well, the
figure is from a photograph of a dried specimen, and repro-
duced here to show the exact length and width of the immature
and lemnisci-bearing segments. Tubercles and mammæform
lemnisci have dried in such a manner that their position is quite
obliterated.

The lemnisci proper are covered by a short tube, and in dry
objects this covering appears to be provided with very small
spines, I mention this because such a spinous integument occurs
only in the species under discussion. and has not yet been observed
in any other Australian cestoid worm. The discharging or male
organs are seldom much produced, they just peep out of the
covering tube, though once I have noticed a lemniscus with a bell

shaped head which is figured (much enlarged) on plate I., No. 15.

I have taken great pains to ascertain how far I am justified in classing so many different-headed cestoid colonies as one species, but as all contain the same kind of ova, (circular bodies enclosing a granular round central capsule without hooks), I must be correct in my observation.

Besides no other cestoid examined was in any way tuberculated. One species closely resembles the present one, and broken pieces of it gave much trouble when classifying the specimens obtained during a day's collecting, but as soon as the test was applied, the different shaped ova proved the fragments to be of another form and not those of *Tænia tuberculata*. I mentioned already that the principal host to this tape-worm is the White-eyed Duck *(Nyroca australis)*.

Many specimens are in the collection of the Australian Museum.

TÆNIA NOVÆ-HOLLANDIÆ.

(Plate I., figs. 1, 1*a*, 1*b*, 1*c*, 1*d*, 1*e*, 2, 3, and 18.
Plate III., figs. 28 and 28*a*).

This species resembles the one previously described and is almost as variable, with regard to the shape of the anterior portion of the colony at least.

Specimens with short tubercular or thin thread-like processes surmounting the broader and thicker " body " or " colony " are common, (figs. 1*b*, 1*d*, and 18 of Plate I.), others occur in which four indistinct suckers can be traced on the most anterior of the segments ; others again resemble the figure given on Plate III., No. 28 and 28*a*, they show four round disks and a retracted proboscis, but all produce the same kind of ova. (Plate I., figs. 2 and 3.) The most common form consists of an oval disk enclosing a tube with round caps at each end and four hooks in the middle. Another ovum occurs with the ends more arched, and the sides bulged out like a cask, (probably a more advanced state) and a third, perhaps an earlier stage, has a double cap on one extremity of the tube (Plate I., fig. 1*c*.) The last was noticed only in one fragment, (Plate I., fig. 1c, nat. size) and its shape may be accounted for, because the ovum was obtained from a very anterior segment; no hooks were seen in it. The present species seldom exceeds 18 inches in length, shorter specimens are, however, of more frequent occurrence. Owing to the lemniscy being long, tapering, and often well thrust out, the sides of a colony appear quite fringed; it is also a curious fact that many of the segments are provided with these organs almost to the very first one. This peculiarity is not shown on Plate I., fig. 1, because I had not noticed it when the sketch was made on account of the lemnisci laying close to the joint, I found, however, afterwards that they always reach right up the sides of the neck, and can be seen with a lens. The joints of the colony are regular and very closely packed for the first inch or two, the last being generally wide and thick. The anterior part in some specimens is very thick also.

Fig. 18 of Plate I. is an enlarged representation of the anterior part of one of the examples which is surmounted by a pennant-like filament. All the specimens obtained were found in the Little Grebe (*Podiceps australis*).

TÆNIA PARADOXA.

(Plate III., figs. 18, 19, 20, 21, 21*a*, 21*b*, 22, 22*a*, and 23).

General form like Tænia novæ-hollandiæ, segments smooth, lemnisci absent. The head is sometimes surmounted by a short proboscis rising between four small distinctly margined suckers or pores, (Plate III., fig. 18.) In other specimens the proboscis is not visible (Plate III., figs. 19 and 20.) At first sight I have often confounded this one with the species which inhabits the intestines of the Little Grebe (*Podiceps australis*), but as the ova differ much in both, I have no doubt that it is distinct.

The ova are round, in fact perfectly circular, covering an oval or half oval body with (in the latter case) a produced smaller half circle in the middle of the less rounded side, (Plate III., figs. 22, 22*a*, and 23.) The hooks are distributed alike in both varieties. The two middle ones vertically and each outer pair in a horizontal position.

Specimens occur from 12 to 18 inches and more in length, and a quarter of an inch wide. Fig. 21*b* of Plate III., represents the posterior segments of the natural size, and fig. 21*a* of the same Plate is a view, (natural size) of the head and neck; the small dot surmounting the figure represents the natural size of 21*a*.

The Little Grebe (*Podiceps australis*) is the only bird in which the *Tænia paradoxa* has hitherto been found.

TÆNIA FORSTERI.

(Plate I., figs. 4, 5, and 6).

Head rather large with four distinct oval suckers or disks, neck distinct, short, closely articulated, the segments being very small but clearly separated. A prominent tubercle on the anterior margin of each disk.

The joints enlarge gradually, their margins are straight, and only the last two or three larger than the rest, the terminal joint being the narrowest in width. Total length two inches and a half.

I have not been able to obtain a single ovum from the few specimens examined, and I find also that the shape of the suckers differs; some heads appear to have them more rounded and the corner tubercle very small. The segments are not furnished with lemnisci.

Habitat : from the stomach of a Dolphin (probably *Delphinus Forsteri*), caught in Port Jackson.

TÆNIA FIMBRIATA.

(Plate I., figs. 22 and 22*a*).

A single fragment about two inches and a half in length and a quarter of an inch wide, is all what I have seen of this singular form. There is no head, and the joints are close together keeping at a uniform distance throughout. The lower margin of each is fringed or rather cut out, in a triangular manner, and the edges are turned over, which gives the fringe a thick appearance. The joints cover each other slightly, and the whole looks like a piece of coarse fibre-matting. The head and neck of this species are missing. Lemnisci occur on both sides, and as the joints are narrow they look very rough and tattered. Mr. Masters who collected the specimen several years ago, was not able to remember the habitat ; it is probably from a Northern Wallaby.

TÆNIA FLAVESCENS.

(Plate 1., figs. 23, 23*a*, 23*b*, and Plate II., fig. 5).

Total length seldom exceeding two inches, head larger than the broadest proglottis, or from $\frac{1}{16}$th to $\frac{1}{8}$th of an inch in diameter.

Head sub-quadrangular, with rounded corners, and bearing four, deep and distinctly margined suckers. Neck segments very close for the length of a quarter of an inch or more, after which they gradually enlarge to the last one, which is smaller The marginal lines are very seldom quite straight, and the edges much produced outwards. Lemnisci were not observed but the posterior half of the joints is furnished each in the centre with a distinct receptacle full of ova. At first these ovaria, if I may call them so, are but scantily filled, but they increase in size and become fuller till the last three or four are completely stuffed out so that the marginal lines between them are altogether obliterated, (Plate II., fig. 5.) The head is broad and rather flat, the neck contracted, and the four circular disks are right upon the upper surface with a small tubercle in the middle. (Plate I., figs. 23, 23*a*, and 23*b*. The disks are not large enough in the sketch).

This is one of the four species in which the colour is beautifully preserved, the proglottides being pencilled and spotted with the most exquisite chrome yellow. I may state that I refer to dry specimens and not to the usual wet preparations ; the process of preservation has been already explained. The Blue Winged Shoveller (*Spatula rhynchotis*), and the Black Duck (*Anas superciliosa*) are the birds in the alimentary canal of which the present species is found. The first one owing to its very narrow and slender intestines is almost destitute of *Tænia*, and whenever these are present, they generally belong to the species just described.

TÆNIA CYLINDRICA.

(Plate II., figs. 6 and 6a).

The head is a little less in size than the broadest proglottis, much produced beyond the neck, subquadrangular with
rounded corners, and with four margined suckers and a tubercular proboscis in the middle. The body consists of closely
packed joints not flattened out but perfectly cylindrical; only
the posterior segments are slightly larger than the others. The
beautiful yellow color so prominent in the allied species described
before has not been noticed in this one, nor can any lemnisci be
traced.

Habitat: the intestines of the Black Duck (*Anas superciliosa*).
Total length half an inch, and one sixteenth of an inch wide.
I have noticed a few much smaller specimens in which the head
was as large as the whole body, but as I could not see any difference except in size, I have arranged them provisionally with the
present species.

TÆNIA CORONATA.

(Plate I., figs. 7 and 8).

Head produced beyond the neck, crown-shaped, with four
large disks and a small proboscis. The neck forms one-fourth of
the total length at which distance the joints begin to increase in
size, the last, which is three lines in width, being the largest.
The lines between the segments are undulated, and appear to be
divided into two or three ridges or elongate tubercles with sometimes an additional one near the openings for the lemnisci, which
are regular and double. Their situation is the upper corner of
each proglottis, they are surrounded by a raised line and scarcely
protrude, so that they are easily overlooked. The lower portion
of each joint bulges out and overshadows the lemniscus of
the next one.

Total length three inches and a half, of which about one inch
forms the thinner part of the neck. The proglottides are rather
wide in a vertical direction, that is in proportion to the size of
the colony; in the middle I counted 12 segments covering a dis-

tance of three lines. No ova were obtained, and there is only a single specimen in the Australian Museum collection.

Habitat: the intestines of the White-headed Stilt *(Himantopus lencocephalus)*.

TÆNIA MASTERSII.

(Plate II., figs. 8, 8a, and 8b).

Head rather small, rounded, not much produced beyond the neck, narrow, with four round and rather indistinct disks. The articulations commence close to the head, gradually enlarging, but not growing much in a vertical direction. Lower margins straight. Four or five deep impressions, forming straight lines, run from the neck to the terminal joints, the outer lines being particularly distinct. Lemnisci were not observed, the specimens appearing all immature, judging from the last rather long and contracted joint; the "Narbe" (scar) of German authors.

Total length from 4 to 5 inches, and about one-eighth of an inch wide in the centre of the colony.

Habitat. The intestines of a species of Wallaby *(Halmaturus)* shot by Mr. George Masters in Queensland.

TÆNIA PHALANGISTÆ.

(Plate II., figs. 7, 7a, 7b, and 7c).

Head, pear-shaped with four large but not very prominent disks upon the upper surface. The disks are flat and not encircled by a ring. Head otherwise granular, and without spines. The neck is considerably contracted at its commencement, and for the first line, (one-twelfth of an inch) does not show any segmentation, it gradually widens out below this point and the joints become distinct. A single specimen is all I possess at present, which is four inches in length. The joints are irregular and distorted, one appears to grow into the other, there are interruptions of the marginal lines, and now and then the joints resemble a series of loops. This state occurs however on the first or anterior half of the colony only. On the remaining portion of the specimen the joints are regular. Some raised dots became visible near the middle of the more posterior segments, but always more on one side than on the other. These dots increase in size as the joints approach the end. No lemnisci have been observed, though many discharged ova were noticed in all directions, some on the side of the head and others on the corners of various joints, (Plate II., figs. 7 and 7a).

Habitat, the intestines of the common Phalanger, (*Phalangista vulpina.*)

TÆNIA PEDIFORMIS.

(Plate II., figs. 2, 2a, 2b, and 2c).

The present species resembles the Hammer-headed Tape Worm of Rudolphi, figured by Bremser, (Icones Helminthum, Plate XV., fig. 17, 18, and 19), and like its European representative inhabits the intestines of various ducks, in this country. Our species are rather larger than the European ones, and more elegant in form, the hammer or foot-like frill is completely articulated, and in very young specimens this part is perfectly erect. It is only when the joints have reached the adult stage, that the "head" or rather the head and neck combined assume the horizontal position with which Helminthologists are most familiar. For the head proper (the first joint or agamozooid) I have looked in vain, and only in one or two instances did I observe two slight sucker-like depressions on the very tip of the thinnest part of the "hammer."

My best and largest specimen was seven inches in length, and occasionally one-eighth of an inch wide; the hammer-shaped combination of head and neck measuring three-eighths of an inch from one extremity to the other. In young specimens the whole, in adults only two-thirds of the appendage, shows articulations, the remaining or lower portions being granular. The joints are very irregular, and occasionally a few indistinct divisional lines appear, but they can never be traced with certainty. There is no sign of any lemnisci, and the edges of the greater portion of the colony are crowded with ova, which during the drying process were copiously discharged. I was unable at the time to sketch them. This species inhabits the intestines of the Black Duck (*Anas superciliosa*), and the Teal (*Anas punctata*).

TÆNIA MOSCHATA.

(Plate II., figs. 9, 9a, 9b, and 9c.)

Head rather prominent, square, with four rounded and produced disks one at each corner. The upper part of the head is divided by a cruciform band with a tubercle in the middle. The segments are not very close together, except on the neck, but are regular in shape, the terminal ones being about a line and a half wide. Total length ten or eleven inches, the greatest breadth being about one-eighth of an inch.

The lemnisci are situated on one side only, near the upper edge of each proglottis, and commence at the distance of half an inch below it, some appear broad and marked with lateral stripes (fig. 9a), others are long and tapering (Plate II., fig. 9c), it is possible, however, that the broad lemnisci are tubes only, the organ being not thrust out. The long and tapering ones occur on the terminal proglottides.

A single perfect specimen taken from the intestines of the Musk duck (*Biziura lobata*) is in the Australian Museum collection. On Plate II. is figured No. 9, the head much enlarged; No. 9b, enlarged proglottides with lemnisci; and 9a and 9c, the organs themselves also enlarged.

TÆNIA RUGOSA.

(Plate II., figs. 4, 4a, 4b, and 4c).

Head small, surmounted by an unarmed proboscis, and now slightly distorted, but fresh probably resembling the head of the Common Duck Tape-worm (Plate III., fig. 4.) The head is provided with the usual four indistinct disks and numerous granular markings, and is attached to a tapering neck which rises suddenly from the base of the more mature segments (Plate II., fig. 4b), and is much contracted just below the head. The proglottides or joints appear mature at the base of the slender neck, they are of moderate size often almost square (fig. 4c), and very rugose (fig. 4, enlarged view.) The lemnisci are on alternate sides, only one to each joint, they are pendulous, short, and covered by a thick sheath. The proglottides resemble those of *Tænia*

o

solium. Total length 5 inches and ⅛th of an inch wide in the broadest part. A single specimen is in the Museum collection which was obtained from the intestines of a White-headed Stilt. (*Himantopus leucocephalus*), shot at the River Hunter, New South Wales.

TÆNIA CHLAMYDERÆ.

(Plate II., figs. 1, 1a, 1b, and 1c).

This is a small species which occurs in the intestines of the Spotted Bower Bird (*Chlamydera maculata*) of this colony. It seldom exceeds three inches in length by a line (one-twelfth of an inch) in width. The head is rounded, flat on the top, and provided with four comparatively large disks or suckers, which in some specimens are separated from each other by grooves. The segments are as usual narrow at the neck, widen out gradually, and show a rather broad marginal line on the upper part with the two lower portions of each joint more or less rugose. Lemnisci could not be discovered. I give a few rough figures on Plate II., No. 1, showing the natural size of the specimen, with an enlarged view of the head, No. 1a, which is rather distorted, the specimen having been crushed, and of some of the joints No. 1b.

TÆNIA BAIRDII.

(Plate III., figs. 1 to 16, and figs. 24, 24a, 26, 27, and 27a).

Looking at the heads of a series of these cestoid worms, it would be quite natural to divide them into at least six species, but as no more than two varieties of ova can be obtained from many supposed species, it is clear that we must look at them as being all identical with one another.

The total length of a mature colony is from three to seven inches, and the broadest posterior proglottides seldom reach the width of one-eighth of an inch. In very few examples lemnisci have been observed, but whenever this was the case, they were noticed to be of great length, from one-third to one-fourth of the width of a proglottis. Many of the mature joints have burst by accident, and in every case the ova were elongate, tube-like or cylindrical in form, the smaller ends slightly rounded off, and containing an S-shaped or roller-like granular body in the centre. On a few occasions four raised lines were noticed in the middle of these bodies, I am unable to say, however, whether they were really spines or not. On Plate III., I have sketched three kinds of ova, but all appear to be identical with one another, as a reference to figs. 2, 6, and 16 will easily show.

After another trial with tape worms the heads of which resemble in shape, figs. 1, 4, 5, 15, and 16a of Plate III., I have obtained nothing but ova as figured under No. 16, granular in the centre and without hooks or spines, I cannot but conclude therefore that my first view was correct, and that all the specimens figured and mentioned above are identical.

With regard to the proboscis it is no doubt retractile, and the hooks may not always be visible, I certainly have seen them occasionally, and have many dried specimens which show them even in that state.

Plate III., fig. 3, is no doubt an accidentally prolonged proboscis, and figs. 8 and 8a may belong to a different species, but I possess only a single dried preparation of it, and therefore cannot apply the ova test. Figs. 9, 10, 11, 12, 13, and 14, are enlarged heads of immature specimens of the same Cestoid; fig. 15a being about the natural size of one, and figs. 16d and 16c show the manner (natural size and enlarged) in which the mature proglottides have burst, no lemnisci being discoverable. Figs. 7 and 7a are representations enlarged and natural size, respectively, of some of the few lemnisci-bearing specimens noticed before, these organs are situated on one side only.

The general form of the *T. Bairdii* is elegant, proportionate, and seldom exceeding seven inches in length, the segments of the " neck " are close together they soon widen out however, but never attain a large size, and of the most posterior proglottides it takes 45, to cover the space of half an inch. The lower margins are generally straight and the edges but slightly bulged out, though now and then more bell-shaped and irregular-margined, proglottides occur.

The head is generally round without, or pear-shaped with, a small proboscis ; four oval disks are indicated but are seldom very distinct, the same may be said with regard to the spines or hooks which in the greater number of specimens are absent or cannot be observed. These tape-worms are numerous in all the Australian Ducks examined, except in the Shoveller, in the Musk Duck and the Pink-eyed Duck. Some of the intestines are almost chocked with them. Smaller cestoid worms occur from a quarter of an inch and less in length, all of which are immature specimens of the present, and probably of one or two other species.

On Plate III. will be found a sketch of these immature bodies which are very difficult to secure in a perfect state, and in most cases are destitute of the head ; I have succeeded, however, in getting a large number of them and will add the following observations. The head generally resembles the mature heads of No. 4 (Plate III.), or the half grown form No. 13, the proboscis being rather large, and to use a popular expression " Top-heavy." specimens resembling fig. 26a of the same plate occur, and many like No. 27a are noticed, either with narrow and long, or with broad and short joints. In the latter case the head has evidently been lost ; I have seen specimens, however, in which the first joint appears quite perfect. It is of no value to discuss every immature variety at present, I therefore draw attention to fig. 25 of Plate III. which may be considered the type of a young specimen of *Tænia Bairdii*, when from a quarter of an inch to half an inch in length. Examples of from one-eighth of an inch to less than a line in length occur in large numbers, all inhabiting, in company with perfectly mature colonies, the same host.

The quantity of young, half grown and mature colonies from a single · water-fowl host is quite astounding, yet all the birds were in excellent condition, and did not appear to have suffered from their guests. On a few occasions I have noticed lemnisci thrust out in young specimens, but have not been able to preserve them, and confess that even with perfectly mature ones this is a rare occurrence. Among perhaps a thousand objects examined, I did not find six with the lemnisci visible. The best specimen is figured natural size, and enlarged on Plate III., figs. 7 and 7a.

Rudolphi illustrates a similar cestoid worm (*Entozoorum*

Plate III.

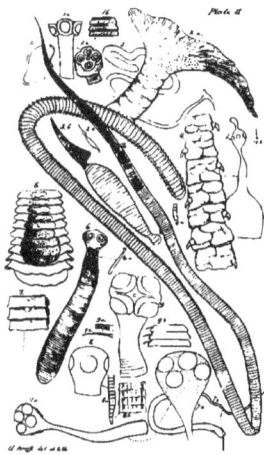

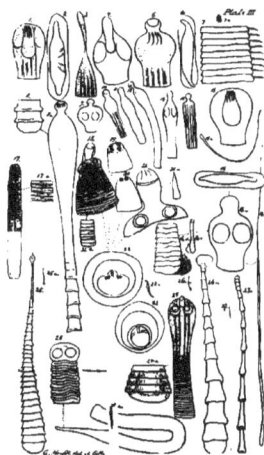

sive vermium intestinalium, &c., Vol. II., Plate I., Tab. X., fig. 2), from the Woodcock (*Scolopax rusticola*) which resembles the young of *Tænia Bairdii*, and is probably also an immature form. This species is found in a great many of the Australian Ducks, in particular in the Black Duck (*Anas superciliosa*.)*

BOTHRIOCEPHALUS (?) MARGINATUS.

(Plate III., figs. 17 and 17*a*).

This species is founded on a fragment about $3\frac{1}{2}$ inches in length, without head, it is fully a line in thickness and nearly two-eighths of an inch wide; joints regular, straight, with very broad and raised margins below, and a slightly raised central papilla. The different segments are not produced outwards, and without the aid of a lens the outer margins appear perfectly straight. Mr. Masters who collected the specimen in some part of Queensland (Wide Bay district I believe) is of opinion that it was taken from the intestines of a Wallaby.

DESCRIPTION OF THE PLATES.

Plate I.

TÆNIA NOVÆ-HOLLANDIÆ.

Fig. 1. —Natural size.

Fig. 1*a*.—A smaller specimen of the same species.

Fig. 1*b*.—Another fragment of the same worm. The more perfect proglottides have parted, and the ovum which is marked fig. 1*c* differs consequently from the more perfect ova.

Fig. 1*d*.—Anterior portion of the specimen No. 1 enlarged.

Fig. 1*e*.—Proglottides, with lemnisci enlarged, taken from the posterior portion of the colony.

* I have just ascertained that the tape-worm figured on Plate III., figs. 24 and 24*a* which I thought was new, is identical (if ova are a test) with *T. Bairdii*. No lemnisci have been seen in this specimen, which was taken from a Stilted Plover.

Fig. 2. —Ovum of the same species with four hooks.
Fig. 3. —A second ovum, probably in a more advanced stage.

TÆNIA FORSTERI.

Fig. 4. —Terminal proglottides enlarged.
Fig. 5. —Outline sketch showing the natural size.
Fig. 6. —Head and neck enlarged.

TÆNIA CORONATA.

Fig. 7. —Head enlarged.
Fig. 8. —Outlines of perfect colony, natural size.

TÆNIA TUBERCULATA.

Fig. 9. —Perfect colony of the natural size.
Fig. 10. —Head and neck, (enlarged) one of the many varieties
          of heads.
Fig. 11. —Head natural size, second variety.
Fig. 12. —Head much enlarged, third variety.
Fig. 13. —Head enlarged, fourth variety.
Fig. 14. —Head enlarged, fifth variety.
Fig. 15. —Lemniscus observed in one specimen.
Fig. 16. —Head much enlarged, sixth variety.
Fig. 17. — Proglottides with lemnisci, enlarged.
Fig. 18. —Head enlarged, eleventh variety.
Fig. 19. — Head much enlarged, seventh variety.
Fig. 20. —Head enlarged, eighth variety.
Fig. 21. —-Head, natural size, ninth variety.
Fig. 21a.—Head, natural size, tenth variety.

TÆNIA FIMBRIATA.

Fig. 22. —Proglottides and lemnisci (enlarged).
Fig. 22a.—Fragment, natural size.

TÆNIA FLAVESCENS.

Fig. 23. —Perfect Colony, natural size.
Fig. 23a.—Head, much enlarged.
Fig. 23b.—Terminal proglottides.

Plate II.

TÆNIA CHLAMYDERÆ.

Fig. 1. —Outline sketch, natural size.
Fig. 1a.—Head enlarged, and slightly distorted.
Fig. 1b.—Segments, enlarged.
Fig. 1c.—Head enlarged, seen from above.

TÆNIA PEDIFORMIS.

Fig. 2. —Outline sketch of the natural size.
Fig. 2a.—Head and neck much enlarged.
Fig. 2b.—Young specimen, much enlarged.
Fig. 2c.—Young specimen, natural size.

TÆNIA TUBERCULATA.

Fig. 3. —Perfect colony, natural size ; from a specimen dried on glass, in which the tubercles had disappeared.

TÆNIA RUGOSA.

Fig. 4. —Proglottides and lemnisci on alternate sides.
Fig. 4a.—Head, neck, and portion of the body of a colony, much enlarged.
Fig. 4b.—Sketch showing the natural size of the same.
Fig. 4c.—Natural size of mature proglottides.

TÆNIA FLACVESCENS.

Fig. 5. —Terminal proglottides of a perfect (mature) colony, showing the heaps of accumulated ova.

TÆNIA CYLINDRICA.

Fig. 6. —Perfect and mature colony (enlarged.)
Fig. 6a.—The same, natural size.

TÆNIA PHALANGISTÆ.

Fig. 7. —Posterior proglottides, with discharged ova and central tubercle.

Fig. 7a.—Outline sketch of head, neck, and portion of body.
Fig. 7b.—Head and neck much enlarged.
Fig. 7c.—Outline drawing of mature and perfect colony, natural
          size.

TÆNIA MASTERSII.

Fig. 8. —Head much enlarged (outline drawing.)
Fig. 8a.—Terminal proglottides slightly larger than natural size.
Fig. 8b.—Proglottides, much enlarged.

TÆNIA MOSCHATA.

Fig. 9. —Head much enlarged.
Fig. 9a.—Lemniscus from the anterior portion of the colony.
Fig. 9b.—Proglottides and lemnisci from the terminal part of
          the colony.
Fig. 9c.—Lemniscus from the last proglottis but one.

Plate III.

TÆNIA BAIRDII.

Fig. 1. —Head enlarged, (variety, No. 1.)
Fig. 2. —Ovum much enlarged.
Fig 3.   —Head and proboscis much enlarged, (variety, No. 2.)
Fig. 4. —Head much enlarged, (variety, No. 3.)
Fig. 5. —Head much enlarged, (variety, No. 4.)
Fig. 6. —Ovum (outline sketch) much enlarged, (variety, No. 2.)
Fig. 7. —Proglottides with their lemnisci, (situated on one side
          only) much enlarged.
Fig. 7a.—Proglottides from the posterior portion of the colony,
          natural size.
Fig. 8. —Terminal proglottides, enlarged.
Fig. 8a.—Head and neck much enlarged, (variety, No. 5.)
Figs. 9, 10, 11, 12, 13, 14—Six varieties of heads of half grown
          and young colonies.
Figs. 15, 15a—Head much enlarged, and outlines (natural size)
          of young colony, (variety, No. 12.)

Fig. 16. —Ovum, (outline.)

Fig. 16a.—Head, much enlarged.

Fig. 16b.—Perfect colony. (Outline sketch of the natural size)
This may be considered the typical form.

Fig. 16c.—Terminal proglottides of the natural size discharging
ova, and without *lemnisci*, though these organs are
sometimes present (see figs. 7 and 7a).

Fig. 16d.—The same view much enlarged.

BOTHRIOCEPHALUS (?) MARGINATUS.

Fig. 17. —Fragment without head (typical specimen).

Fig. 17a.—Proglottides enlarged, shewing the central pore or
orifice.

TÆNIA PARADOXA.

Fig. 18. —Head, side view much enlarged (variety No. 1).

Fig. 19. —Head, upper view shewing the four suckers, much
enlarged.

Fig. 20. —Head side view shewing only two suckers.

Fig. 21. —Head, with proboscis and suckers much enlarged
(outline sketch).

Fig. 21a.—The same head, natural size, variety No. 3. The
small dot surmounting the conical neck represents
the natural size of the much enlarged figure
No. 21.

Fig. 21b.—Terminal proglottides, natural size.

Fig. 22. —Ovum, with six-hooked embryo.

Fig. 22a.—One of the hooks ; rough sketch (enlarged).

Fig. 23. Ovum, with six-hooked embryo, a second stage, and the
most common form observed.

TÆNIA BAIRDII.

Fig. 24. —Head and neck much enlarged, variety No. 14.

Fig. 24a.—Terminal proglottides shewing accumulated ova, but
no lemnisci.

Fig. 25. —Young colony, much enlarged (the most common form).

Fig. 25*a*.—The same colony, natural size.

Fig. 26. —A second variety of a young colony, of the natural size.

Fig. 26*a*.—The same, much enlarged.

Fig. 27. —A third variety of a young colony of the natural size.

Fig. 27*a*.—The same, much enlarged.

TÆNIA NOVÆ-HOLLANDIÆ.

Fig. 28. —Head, much enlarged.

Fig. 28*a*.—Perfect colony, outline sketch, natural size.

P.S.—It may be of interest to know that the number of fowls, constantly feeding on some of the Tænia which escaped during the cleaning process, did not suffer in consequence. Some killed at various times, by way of experiment, were totally free from worms.

Descriptions of eight new species of Stephanopis (Cambridge), by

H. H. BURTON BRADLEY, ESQ.

[Read 7th August, 1871.]

THE genus *Stephanopis* was founded by the Rev. O. P. Cambridge on specimens described by him in the Annals and Magazine of Natural History, for January 1869, and was by him hesitatingly placed among the *Thomisides;* the further species described below with the knowledge of their habits enable me to add my opinion to his as to the placing of the genus—at least, so far as regards the four first species described by him *(S. altifrons, nigra, clavata, lata),* and those which I have described hereafter. To the characters of the genus given by Mr. Cambridge, I should add—

Legs : certainly laterigrade.

Habits : living under loose bark of trees without tube or cell of silk.

STEPHANOPIS CAMBRIDGEI.

♀ length 4 lines.

Cephalothorax : about 1½ lines long ; breadth a little less. *Caput* and *Clypeus* as in *S. altifrons,* but the cephalic protuberance is more developed than in that species ; color above, of a greyish-brown, much darker at the edges ; the tip of the cephalic protuberance greyish-white.

Clypeus : light grey.

Eyes : anterior intermediate half the size of the others, which are nearly equal ; the anterior laterals being slightly largest ; all about equidistant, but anterior intermediates are placed much below the others.

Legs : moderately long, very nearly equal ; two anterior pairs slightly longest and stronger than the two posterior pairs ; the legs are otherwise as in *S. altifrons.* Color, 1st and 2nd coxæ ; yellowish-brown ; femur, nearly black with greyish patch on upper side ; tibia, not so dark ; tarsi and metatarsi, greyish-brown, banded with stripes of a darker shade ; 3rd and 4th coxæ ; and femur, yellowish-brown with darker spots ; the other parts greyish-brown, banded.

Palpi : moderate in length and strength ; color and armature as in tarsi of legs.

Falces : moderately long, strong, brownish-red, with yellowish spots, hairy, inclined backwards towards the *maxillæ.*

Maxillæ : reddish, but lighter on the inner edges, moderately long, nearly straight ; rounded on the outer sides at the apex ; inclined towards the labium which is of the same color as maxillæ, about as broad as long, slightly rounded at the apex, and broadest one-fourth of its length from base.

Sternum : large cordate, narrowest at its posterior part, slightly rugulose and hairy, of a yellowish brown color, hair light grey.

Abdomen : depressed, about 2½ lines long, breadth a little less. posterior broadest and thickly furnished with bluntish tubercu-lated spines and bristles similar to those on legs ; color above, same as *cephalothorax,* but having a dark longitudinal mark in the middle ; the edges are greyish, the sides dark with two darker transverse markings at posterior part ; below—rugulose at the sides, centre part defined, less rugulose and of a darker color.

This insect is from Nepean Towers, where I obtained it in August 1870.

S. MONTICOLA.

♀ length a little over 4 lines.

Cephalothorax : nearly two lines long, not quite so broad ; cordate, broadest at the posterior part, in shape otherwise as in *S. altifrons,* but not so long in proportion ; color above, greyish with brown markings, caput reddish brown. *Clypeus,* greyish but lighter ; the whole slightly rugulose.

Eyes : as in *S. Cambridgei,* but the cephalic protuberance is not so developed.

Legs : relative length as in *S. Cambridgei* ; color above as *cephalothorax,* but there are no regular bandings ; tuberculate with blunt spines ; below the coxa and femur, yellowish.

Palpi : moderate in length, strong ; color, as legs.

Falces : same color as *clypeus* but a shade lighter, shape as in *S. Cambridgei.*

Maxillæ and *labium :* form as in that species, but the color is a reddish-brown.

Sternum: moderate size, oval, approaching cordate; broadest at base; color, yellowish.

Abdomen: depressed, nearly $2\frac{1}{2}$ lines long, and as broad at posterior part, which is broadest, and tuberculated on upper sides; color above, as *cephalothorax*, darker in the centre and towards the sides; below, yellowish-brown in the centre; yellowish-grey towards the edges; the sides are greyish-brown; the two posterior spinnerets slightly prolonged.

This insect is altogether not so strongly tuberculated as *S. Cambridgei.* I obtained it in February 1868, at Tia, New England, under bark, without either web or tube of silk. Tia is on the coast side and almost on top of the coast range; I am told about 3000 feet above the level of the sea.

S. TUBERCULATA.

♀ length a little over $3\frac{1}{2}$ lines.

Cephalothorax: black rugulose, nearly two lines long, about one broad at posterior part, which is slightly broadest and narrowing gradually to base of cephalic protuberance, which is narrow and rises abruptly.

Eyes: placed much as in *S. nigra*, but lateral anteriors nearly twice as large as posterior eyes, which again are about twice as large as intermediate anteriors.

Legs: 1st and 2nd equal, longest; 3rd and 4th equal; all strong and moderately long; 1st and 2nd very strong in the femur and tibia, and tuberculate; 3rd and 4th less so; color, above black; below, as far as the tarsi, yellowish-brown, tarsi and metatarsi, blackish-brown.

Palpi: yellowish-brown, strongly armed with tubercles and blunt spines, which are nearly black.

Falces: shape as in *S. altifrons;* color, yellowish-grey; fangs, reddish, moderately long, slightly curved.

Maxillæ: shape as in that species; color, same as that of falces; *labium*, about as long as broad, and rounded at the apex; broadest at the base.

Sternum: broad, oval, truncate at fore part, a few light colored hairs scattered over it; color, same as maxillæ, but slightly darker towards the edges.

Abdomen : depressed, nearly two lines long, about one line broad, nearly oval ; above, black rugulose, tuberculated towards the base ; fore part notched, and rising above base of *cephalothorax ;* below, of a yellowish-brown.

This insect also caught at Nepean Towers ; was found under bark without web or tube of silk, in December 1866.

S. DEPRESSA.

♀ length nearly 5 lines.

This insect, in the form of all its parts, closely resembles the preceding.

The *cephalothorax :* a little over two lines long and broad, is of a reddish-grey, with two bands of reddish-brown meeting at the base of the *caput,* curving outwards from that point.

Clypeus : reddish-brown.

Legs and *Palpi :* yellowish-brown ; the tubercles of brown, which give a mottled appearance above.

Falces : yellowish-brown.

Maxillæ and *Sternum :* reddish-brown, with a few scattered hairs of a light color.

Abdomen : depressed, nearly three lines long and nearly as broad at posterior part, which is broadest ; color, greyish-brown with nearly white patches, tuberculated slightly to the sides, which are slightly darker ; below, of a dirty brown.

I have this insect from Cape York, N.A.

S. ELONGATA.

♀ adult ? length 3½ lines.

This insect differs from the preceding, which it greatly resembles both in form and color ; in the shape of the *cephalothorax* which is round ; the legs are longer in proportion ; the whole insect is flatter, and more tuberculated at posterior part.

Sternum : is broadest in rear, truncate in front ; the *femur* and *tibia* of first pair of legs are of a darker color ; the distinct marking on the *cephalothorax* of the preceding species does not appear in this one, but there are two tubercles of a darker color in the centre of the *cephalothorax ;* the general color of this species is the same but lighter than that of the preceding.

I have this insect also from Cape York.

S. THOMISOIDES.

♀ length a little over 3 lines.

Cephalothorax: reddish brown, 1 line long, 1 wide at base, cordate—cephalic protuberance not very high, slightly rugulose.

Clypeus: not cleft, slightly prominent, lighter shade of same color.

Eyes: as in *S. tuberculata*.

Legs and *Palpi*: same color as *cephalothorax*; armature and relative length as in *S. tuberculata*.

Falces: same shape as in that species, lighter red, hairy at tips; fangs red, moderately long and curved.

Maxillæ: straight on outer side, moderately long, not much inclined on *labium*, rounded on inner side.

Labium: broader than long; rounded at apex.

Sternum: as *S. tuberculata*; color, reddish-brown.

Abdomen: depressed; slightly cleft in front; about two lines long; broadest at posterior part, and broader than long; above, lighter shade of same color as *cephalothorax* with dark band running along sides; slightly tuberculated at the rear; below, same color but lighter.

This insect, also from Cape York, in general appearance closely approaches the genus *Thomisus*.

S. RUFIVENTRIS.

♀ length 5 lines.

Cephalothorax: 1½ lines long and as broad; dark reddish-brown; cephalic protuberance moderately high and not cleft.

Clypeus: not very prominent; tuberculate red; the cephalic protuberance is continued by a gradual distinct fall to the posterior part of the *cephalothorax*.

Eyes: anterior intermediate extremely small and not much lower than laterals; anterior laterals twice as large as posteriors; posterior laterals placed lower than intermediates.

Legs: as in *S. altifrons*; color, uniform dark reddish-brown.

Palpi: moderate in length and strength; similar to legs in color and armature.

Falces: long and strong; similar in color to legs; much inclined backwards.

Maxillæ and *labium :* as in *S. altifrons* ; color, light red.

Sternum : large ; nearly round ; clothed with short hairs ; same color as labium.

Abdomen : above, yellowish red with dark red-brown bandings " en chevron ; " slightly rugulose, and furnished with short blunt spines more than three lines long and about as broad, oval, slight, truncated at the front ; below, reddish-yellow.

I obtained this insect under bark, without web or tube of silk, at Tia, in February 1868.

S. MACLEAYI.

♀ length 5 lines.

Cephalothorax : nearly two lines long and not quite as broad, and otherwise very like *S. rufiventris ;* the cephalic protuberance and clypeus also as in that species.

Eyes : as in that species.

Legs : very much similar to those of that species, but the femur of the first pair is more strongly developed ; color, dark reddish brown with irregular brown blotches on femur, two yellow bands on tarsi of third and fourth pairs.

Palpi : moderate in length and strength, similar to the legs in color and armature.

Falces : as in *S. altifrons ;* color red, the fangs curved and surrounded with thick reddish hairs.

Maxillæ : moderately long, hollowed on the inner side, rounded slightly on the outer, and surrounding the labium which is as in *S. altifrons.*

Sternum : large oval, narrowest at its fore extremity ; clothed with hairs ; reddish brown, with a longitudinal hollow which is darker in the centre.

Abdomen : above 3 lines long and about $2\frac{1}{2}$ broad, oval, reddish-brown, a light band extending along the edges ; moderately rugulose, hinder part furnished with tubercles and blunt spines ; below, yellowish-brown.

The insect I have described above I received from W. MacLeay, Esq., after whom I have named it. I believe he obtained it near Goulburn.

Notes on a collection of Insects from Gayndah, by

WILLIAM MACLEAY, ESQ., F.L.S.

SECOND PAPER.

[Read 4th December, 1871.]

I NOW proceed with the list of Gayndah Coleoptera, in continuation of my paper of last April. I was in hopes at one time of having been able to bring my task to a termination shortly, but press of business and other matters have reluctantly compelled me to give up all idea of proceeding for some time at least beyond the limits of the present paper.

BUPRESTIDÆ.

420.—NASCIO VIRIDIS. n. sp.

Length 4 lines.

Elongate, narrow, green, opaque, and punctate. Thorax slightly lobed in front, truncate behind, and parallel sided, with a transverse impression near the base, and a small round fovea in the centre of the base. Elytra coarsely striato-punctate—the interstices near the suture elevated,—a little broader than the thorax at the base, and wedge shaped and serrate towards the apex which is bidentate, with the posterior two thirds of the suture of a bluish tinge and a small yellow spot in the centre of each elytron. Body beneath and legs green, subnitid, punctate, and thinly clothed with a whitish pubescence.

421.—ASTRÆUS MASTERSII. n. sp.

Length 4 lines.

Head punctate, whitish pubescent, black on the occiput, green in front, with a slight central longitudinal ridge. Thorax punctate, slightly pubescent, and black with the sides green. Elytra black, tinged with green and purple, striato-punctate—the in-

P

terstices elevated,—and marked with a large round spot near the
base, a fascia above the middle extending on the sides to near
the humeral angle and not reaching the suture, another fascia
behind the middle also not reaching the suture, and a small round
spot near the apex, all of a golden yellow.   Under side of body
brassy green, with purplish spots on the abdomen.   Legs green,
with the apex of the tibiæ and the tarsi yellow.

422.—MELOBASIS AZUREIPENNIS.   n. sp.

Length 4½ lines.

Head blue, densely clothed with golden hair.   Thorax of a
fiery copper colour and punctate, densely on the sides, thinly in
the middle.   Elytra blue, nitid, striato-punctate—the interstices
smooth,—and strongly serrated at and near the apex.   Body
beneath brassy green, changing to steel blue towards the apex of
the abdomen.   In the female the under surface and legs are
entirely blue.

I find specimens from Port Denison of this beautiful species
in my collection labelled *M. azureipennis* La Ferte, but I have not
been able to find any notice of such an insect having been
described.

423.—MELOBASIS COSTATA.   n. sp.

Length 6½ lines.

Elongate, bronze green, subopaque, densely and finely punctate.   Head clothed with whitish pubescence.   Thorax less
densely punctate than the rest of the body, with a minute fovea
in the centre of the base, and with an almost imperceptible
smooth space marking the median line.   Scutellum small, rounded,
green, smooth, and nitid.   Elytra slightly serrated towards the
apex, and marked with three elevated smooth lines, the inner
one largest and nearly reaching the apex, the second less distinct but nearly as long, the third least distinct and much shorter.
There is also a short costa near the scutellum extending from
the base to the suture at about one third of its length.   Legs
coppery red, tarsi cyaneous.

424.—MELOBASIS APICALIS. n. sp.

Length 4½ lines.

Green, subnitid, densely punctate. Head brassy in front and roughly punctate. Thorax transversely punctate, with a short clear space marking the base of the median line. Elytra transversely depressed near the base—the depression not extending to the suture—and serrate towards the apex, with the apex and the apical portions of the suture and sides of a bluish purple. Body beneath and legs of a coppery hue, clothed with a short white pubescence. Tarsi green.

425.—MELOBASIS OBSCURA. n. sp.

Length 2½ lines.

Reddish brown, opaque, densely punctate, and of an oblong oval form. Thorax much broader than the length. broader behind than in front, slightly bisinuate at the base and transversely punctate. Scutellum small, oval, depressed and punctate. Elytra transversely impressed near the base, rather flat, of the width of the base of the thorax at the base, minutely serrated behind, separately rounded at the apex, and clouded with some dull coppery red patches

426.—NEOCURIS MASTERSII. n. sp.

Length 2¾ lines.

Subelongate, black, subnitid, punctate. Head deeply impressed between the eyes and of a bluish tinge. Thorax subconvex, and roundly lobed at the base. Scutellum small, subtriangular, punctate. Elytra purplish black, striato-punctate—the interstices elevated—with a small yellow spot at the base of the fifth interstice, another on the lateral margin, and a large round spot of the same colour above the middle of each elytron. The legs, antennæ, and underside of body are cyaneous.

427.—NEOCURIS GRACILIS. n. sp.

Length 1½ lines.

Greenish black, subnitid, punctate. Head green in front and without frontal impression. Thorax slightly lobed behind, and minutely foveate in the centre of the base, with the sides of a fiery copper colour. Scutellum small, subglobular and smooth. Elytra transversely rugose, with the apex separately rounded and somewhat dehiscent. Legs viridi-æneous.

428.—Anthaxia obscura. n. sp.

Length 2¼ lines.

Brownish or greenish black, opaque, subgranulate, shallowly and densely punctate. Forehead viridi-æneous, clothed with white hair. Thorax twice as broad as the length, truncate behind, and emarginate on the sides near the posterior angles. Scutellum small, elongate, subtriangular, not pointed. Elytra of the width of the thorax at the base, and separately rounded at the apex, with a distinct transverse line at the base, some shallow indistinct depressions near the base, and a fine stria on each side of the suture.

t.

429.—Anthaxia cupripes. n. sp.

Length 4 lines.

Greenish black, subopaque, punctate. Forehead clothed with whitish pubescence. Thorax nearly twice as broad as the length, roundly and broadly lobed in the middle of the base, with a small fovea marking the base of the median line. Scutellum small, cordate, pointed at the apex and punctate. Elytra transversely marked at the base, finely serrated on the sides posteriorly, and rather acutely roundly at the apex. Under side of body very brilliant, green in the centre, and cupreous on the sides and apex. Legs of a coppery red.

430.—Anthaxia purpureicollis. n. sp.

Length 2¾ lines.

Black, subopaque, punctate. Forehead clothed with white pubescence. Thorax purple, subnitid, lobed at the base, with a minute fovea in the extremity of the lobe. Scutellum subtriangular, punctate. Elytra bisinuate at the base, transversely depressed near the base, acutely rounded and dehiscent at the apex, and very minutely serrated on the sides towards the extremity. Body beneath and legs brassy, punctate and subnitid.

431.—ANTHAXIA NIGRA. n. sp.

Length 2¼ lines.

Differs from the last in being much smaller, in being entirely black, in having the median line of the thorax slightly marked in front, and in having the scutellum raised in the centre.

The three last species agree in having the base of the thorax roundly lobed in the centre and sharply at the posterior angles, and in so far differ materially from the typical form of Anthaxia.

NOTOGRAPTUS. n. gen.

This genus seems to be nearly allied to *Anthaxia*. The antennæ are identical, the epistome, the head, the eyes, the palpi, the position of the antennal pores, the form of the antennal cavities, the prosternum, metasternum and legs likewise accord in almost every particular. The labrum, however, is rounded at the apex, the thorax is transverse, rounded on the sides, considerably narrowed at the posterior angles, and bisinuate at the base with a central broadly rounded lobe. The scutellum is small, of the form of an equilateral triangle, and depressed on the surface. The elytra are broad and rather flat.

432.—NOTOGRAPTUS SULCIPENNIS. n. sp.

Length 4¾ lines.

Brownish black, very opaque, densely punctate. Forehead clothed with white hair, and with two small tubercles between the eyes. Thorax with three broad longitudinal depressions clothed with silvery hair, the central one being on the median line. Elytra as broad as the thorax at its broadest part, rounded at the humeral angles, and conjointly rounded at the apex, with two broad longitudinal depressions on each elytron. one extending from the humeral angle downwards and inwards for two-thirds of its length, the other outside of the first and short. Under side of body and legs black and subnitid.

433.—NOTOGRAPTUS HIEROGLYPHICUS. n. sp

Length 2¼ lines.

This insect is very much like the last species, it differs chiefly in its much smaller size, in the head being without tubercles and with the white hair forming two rather indistinct fasciæ, in the thorax not having the lateral longitudinal depressions extending to the apex, and in the elytra having a rather narrow line of golden pubescence extending from the base to near the apex, in a series of zigzag and rectangular forms.

434.—CURIS SPLENDENS. n. sp.

Length 6 lines.

Head brassy green, densely punctate, and deeply excavated in front along the median line. Thorax broader than the length, a little broader behind than in front, somewhat lobed in the middle at the apex and base, rounded at the anterior angles, acute at the posterior angles, and of a dark blue colour, with the median line and the sides of a reddish golden lustre and punctate, the former narrow in front and broad and foveate behind, the latter broad and but slightly foveate. Scutellum small, round, and of a reddish golden hue. Elytra as broad as the thorax at the base, slightly narrowed behind, obliquely truncate at the apex, slightly sinuate on the sides, irregularly and coarsely punctate, of a coppery red on the suture and sides, with a strong costa near the sides behind and with two costæ on the middle of each elytron which portion is of a greenish hue. Body beneath brassy green, legs blue. The male is smaller and of less brilliant colouring than the female.

The most marked distinguishing feature in this species seems to be the broad obliquely truncated elytra. I have seen several species from northern parts of Australia similar to it in this respect, but none of them have been described.

435.—STIGMODERA IMPRESSICOLLIS, MacL., W. Trans. Ent. Soc. N. S. Wales, I., page 32.

436.—STIGMODERA SEXGUTTATA, MacL., W. Trans. Ent. Soc. N. S. Wales, I., page 29.

437.—STIGMODERA DISTINCTA, Saund. *Journ. Linn. Soc.*, 1868, *page* 473.

438.—STIGMODERA MASTERSII. n. sp.

Length 5 lines.

Brassy black, subnitid, finely punctate. Forehead broadly excavated. Thorax foveate in the centre of the base. Scutellum transverse, smooth, pointed and of a greenish black. Elytra striato-punctate, obliquely truncate and strongly bispinose at the apex, and of a yellowish red colour with a spot adjoining the scutellum, a narrow fascia behind the middle and a large transverse spot reaching the suture near the apex, of a bluish black. Under surface and legs bluish black, subnitid.

This species is somewhat like the *S. distincta* of Saunders and *S. Andersonii* of Laporte and Gory.

439.—STIGMODERA VIOLACEA, MacL., W. *Trans Ent. Soc. N. S. Wales*, I., *page* 23.

440.—STIGMODERA RUFIPES, MacL., W. *Trans. Ent. Soc. N. S. Wales*, I., *page* 23.

441.—STIGMODERA KREFFTII. n. sp.

Length 4 lines.

Head and thorax brassy black, punctate and subopaque, the latter lightly marked on the median line and foveate at the base. Scutellum green, coarsely punctate. Elytra punctato-striate with the second interstice strongly costate, minutely bidentate at the apex, and of a blackish purple colour, with a large spot at the base not touching the suture, and extending along the side at the humeral angle, a broad median fascia not reaching the suture, and a smaller subapical fascia, also not reaching the suture, all of a reddish yellow. The subapical fascia is extended a little along the lateral margin of the elytra and is there of a deep red colour. Under side of body brassy black, and slightly pubescent. Legs and parts of the mouth dark blue.

442.—Stigmodera elongatula.   n. sp.

Length 4½ lines.

Elongate, narrow, subnitid, punctate.   Head and thorax brassy green, the former very slightly excavated in front, the latter foveate and squamose at the sides, and slightly depressed at the base, with the median line lightly marked.   Scutellum rather elongate, depressed, not punctate.   Elytra punctato-striate,—the second interstice larger than the others,—closely bispinose at the apex, and of a purplish or bluish black colour, with four discal and two lateral yellow spots on each elytron placed much in the same way as in *S. Xanthopilosa* Hope. Under side of body nitid, with silvery pile or scales.

I have in my cabinet a Sydney species, which I believe to be unnamed, which only differs from this insect in being of a more brilliant colouring and in having the scutellum punctate.

443.—Polycesta Mastersii.   n. sp.

Length 9½ lines.

Brown, opaque, rough, and somewhat flat.   Head broadly depressed in front.   Thorax much broader than the head, nearly truncate in front, rounded on the sides, evidently broader behind than in front, slightly sinuate at the base, foveolate in the middle, and marked on the median line by a small elevation.   Scutellum small, truncate, black and nitid.   Elytra coarsely striato-punctate and somewhat pointed at the apex.   Under side of body with a slight coppery reflexion.

444.—Chrysobothris Saundersii.   n. sp.

Length 7½ lines.

Of a somewhat nitid bronzy colour, densely and finely punctate.   Head pilose in front, coppery red and carinate on the summit.   Thorax transversely punctate, and obliquely bifoveate near the sides, with the median line lightly marked.   Scutellum small, triangular, and smooth.   Elytra flat, and serrate behind, with a fine costa near the suture, an abbreviated one about the middle, a fovea near the humeral angle, and three small coppery red foveæ on each elytron, one at the base of the first costa, the second on the middle of the second costa, the other between that and the apex and outside the second costa. Body beneath brassy black, nitid and thinly clothed with a whitish pubescence.

445.—CHRYSOBOTHRIS MASTERSII.   n. sp.

Length 6 lines.

This species differs from the last in being less flat, in having the head less strongly carinate on the summit, in having the thorax of a more purple hue, the lateral foveæ scarcely visible, the median line not traceable, and the basal lobe more rounded, in having the elytra of a bluish black colour, without the humeral fovea, with the costæ less marked, and with the three discal foveæ larger and of a golden lustre, and in having the under side of the body very brilliant.

446.—CHRYSOBOTHRIS VIRIDIS.   n. sp.

Length 5 lines.

Dark bronzy green, nitid, and finely punctuate. Head flat in front, thickly clothed with a short silvery pubescence, and not carinated on the summit. Thorax transversely punctulate and shaped much as in the last described species. Scutellum somewhat transverse and pointed behind. Elytra flatter than in the last species, with the first and second costæ as in *C. Saundersii*, with a third costa distinct near the apex, with a deep humeral fovea, and with the three discal foveæ large, placed as in the last species and of a golden green colour. Under side brilliant and green in the centre, with the sides and legs cupreous.

447.—ETHON LATIPENNIS.   n. sp.

Length 5 lines.

Ovate, transversely punctate, subnitid, squamose, and of a black colour. Head golden and covered with whitish hair, which forms a prominent tuft near each eye. Thorax rough, foveate at the sides, smoother in the middle, distinctly marked and finely punctate on the median line, and having a metallic gloss on the margins and elevated roughnesses. Scutellum viridi-æneous, almost smooth. Elytra broader than the thorax, subdepressed, parallel-sided for two-thirds of their length, narrowed on the apical third and separately rounded at the apex, with a strong tubercle near the humeral angle, an elongate one between that and the scutellum, and with numerous patches of short black scales forming towards the apex an irregular fascia. Under side of body and legs' dark blue, and finely punctate, with the apex of the abdomen and under side of the tibiæ, cupreous.

448.—CISSEIS DIMIDIATA. n. sp.

Length 3½ lines.

Entirely of a metallic green, excepting the apical half of the elytra and more than half the suture, which are of a coppery red, Head densely punctate, and lightly excavated in front. Thorax transversely punctate and without foveæ or median line. Under side of body finely punctate.

449.—CISSEIS IMPRESSICOLLIS. n. sp.

Length 4 lines.

Green, nitid, punctate. Thorax impressed longitudinally at the base near each side, the impression extending to the elytra. Scutellum scarcely punctate. Elytra rather elongate and of a purplish green with several round shallow foveæ of a brassy green. Body beneath very nitid.

450.—CISSEIS VIRIDI-AUREA. n. sp.

Length 4 lines.

This species closely resembles the last in form and sculpture. It is however entirely of a brilliant golden green, and the shallow foveæ on the elytra are larger, covered with short white setæ or scales and are without metallic gloss.

451.—CORAEBUS MARMORATUS. n. sp.

Length 3¾ lines.

Brassy black, nitid, densely punctate. Head golden coloured and lightly excavated. Thorax rounded at the posterior angles, broadly rounded at the basal lobe, and thickly clothed except in the middle with short whitish scales. Elytra of the width of the thorax at the base, becoming narrower on the posterior third, separately rounded at the apex, and marbled all over with patches and fasciæ of short whitish scales. Body beneath bluish black.

452.—AGRILUS MASTERSII. ·n. sp.

Length 6 lines.

Dark brown, opaque, finely punctate. Head of a roseate hue in front and lightly foveated. Thorax deeply marked on the median line except in front, and with a short sublateral elevated line and a fovea in front of it. Elytra largely foveate at the base, unicostate, and pointed and serrate at the apex. Body beneath cupreous and nitid, with a short silvery pubescence.

453.—AGRILUS DEAURATUS. n. sp.

Length 3 lines.

Very narrow, black, subnitid with an occasional metallic gloss, and finely punctate. Head depressed on the vertex. Thorax transversely punctate, deeply marked on the median line behind, and rather longer than the width. Scutellum transversely carinated. Elytra bispinose at the apex, and of a tarnished looking golden yellow colour, with an oblique dark blue fascia near the apex. Body beneath blue and thinly punctate.

EUCNEMIDÆ.

Of this family there are three species in the collection, all of different genera, and I believe undescribed, but as I have not been able to procure Bonvouloir's Monograph of the group, in which he adds many genera and species to those previously known, I must pass them over for a time.

ELATERIDÆ.

454.—AGRYPNUS MASTERSII. n. sp.

Length 15 lines.

Brown, subopaque, finely punctate and clothed with a silky ashen pubescence. Antennæ and palpi dull red. Head lightly excavated in front. Thorax longer than the width, narrower in front than behind and rounded a little on the sides, with the anterior angles prominent, the posterior divergent subacute and prolonged backwards, the median line slightly marked near the base where it bisects the central tubercle, and with the sides finely carinated along the basal half. Scutellum large, longer than the width, and broadly rounded at the apex. Elytra as broad as the thorax at the base, sinuate behind the humeral angles, narrowed gradually and rounded at the apex with eight distinctly punctured striæ on each elytron. Legs clothed with a very silky pubescence.

455.—AGRYPNUS LATIOR. n. sp.

Length 16½ lines.

This species is longer and broader than the last, and differs from it also in having the antennæ and palpi brown, the basal tubercle of the thorax transverse and not bisected by the median line, the posterior angles shorter and more rounded though still pointed backwards, and the lateral carination extending beyond the posterior half. The scutellum differs also in having three impressions on the posterior margin. In this as in the former species the striæ on the elytra are most deeply punctate towards the sides.

456.—LACON MAMILLATUS, Cand. *Mon.* 1, *page* 107, *t.* 2. *f.* 3.

457.—LACON GAYNDAHENSIS. n. sp.

Length 6½ lines.

Dark brown, subopaque, punctate—each puncture furnished with a yellow decumbent setiform scale. Thorax subconvex, longer than the width, rounded on the sides in front, slightly narrowed towards the base, and widened again at the posterior angles which are acute, with the base largely but not deeply emarginate, the median line scarcely traceable, and the lateral carination very small and extending along the basal half. Elytra narrowed and rounded at the apex, and striato-punctate—the punctures large and quadrangular.

458.—LACON ALTERNANS. n. sp.

Length 4¼ lines.

Reddish brown, opaque, elongate, flat, punctate and clothed with cinereous scales. Head slightly depressed in the middle. Thorax longer than the width, rounded on the sides and almost truncate behind, with the posterior angles curved outwards, rather rounded, and without carination. Elytra subacutely rounded at the apex, and covered with rows of large punctures, with every second interstice subelevated and densely clothed with scales.

459 —LACON MACULATUS. n. sp.

Length 3¾ lines.

Brown, opaque, flat, densely punctate, and clothed with very short setiform scales of a brown colour, interspersed with numerous indistinct spots of reddish brown and cinereous scales. Thorax with the median line traceable, the posterior angles curving slightly outwards and truncate at the apex, and the base rather deeply bi-emarginate. Scutullum nearly round. Elytra strongly striato-punctate, slightly widened about the middle, and narrowed and rounded at the apex. Antennæ and tarsi pale red.

460.—LACON GRANULATUS. n. sp.

Length 3 lines.

Dark brown, opaque, broad, flat, densely punctate, granulose looking, and scaly. Thorax much longer than the width, with the posterior angles not externally produced, and with the base deeply bi-emarginate. Scutellum subtransverse. Elytra strongly bisinuate at the base, not much longer than the width, narrowed roundly at the apex, angularly sinuated behind the humeral angles, and striato-punctate with the alternate interstices subcostate. Beneath with the extremities of the prothorax, the lateral margins of the abdominal segments, and the legs, piceous red.

461.—APHILEUS LUCANOIDES, Cand. Mon. 1, page 184, t. 3. f. 5.

462.—MONOCREPIDIUS MASTERSII. n. sp.

Length 10 lines.

Dark brown, subopaque, finely and densely punctate, and clothed with reddish yellow decumbent pile. Antennæ and palpi reddish. Thorax subconvex, scarcely longer than the width, and very lightly marked on the median line near the base, with the posterior angles strong, bicarinate, and rather obtuse. Scutellum more elongate than in *Monocrepidius Australasiæ*. Elytra scarcely longer than twice the width, broadest about the middle, slightly narrowed and rounded at the apex, and strongly striato-punctate. Legs reddish.

463.—MONOCREPIDIUS STRIATUS. n. sp.

Length 8 lines.

Of a more elongate form than the last species, brown, subopaque, very finely and densely punctate, and thickly clothed with a pale fulvous decumbent pile. Thorax longer than the width, with the median line marked on the posterior half, and with the posterior angles strong, subacute, and bicarinate—the inner carina small. Elytra nearly three times as long as the width, striated—the interstices nearly flat and densely punctate,—and separately subacuminate and minutely emarginate at the apex. Legs and antennæ yellowish red.

464.—MONOCREPIDIUS ACUMINATUS. n. sp.

Length 6 lines.

Subelongate, reddish brown, subopaque, very finely punctate, and clothed with a very short ashen pile. Head somewhat depressed in the middle. Thorax much longer than the width, with the median line scarcely traceable behind and the posterior angles strong, subacute and bicarinate—the inner carina very small. Elytra finely punctato-striate, with two reddish patches at the base, and with the apex emarginate and minutely toothed at the outer extremity of the emargination. Legs yellow.

465.—MONOCREPIDIUS BREVICEPS. n. sp.

Length 7 lines.

Subelongate, brown, subopaque, finely punctate, and densely

clothed with a pale fulvous pubescence. Top of the head as far as the frontal ridge short and horizontal. Thorax elongate, with the median line distinct throughout, and the posterior angles acute and bicarinate. Elytra reddish at the base, nearly three times as long as wide, narrowed and separately rounded at the apex, and striato-punctate, with the interstices nearly flat. Antennæ and the parts of the mouth red. Legs reddish brown and silky.

466.—MONOCREPIDIUS RUBICUNDUS. n. sp.

Length 5½ lines.

This species is of a less elongate form than the last, redder in colour and more nitid. The head is less horizontal and short on the upper part. The thorax is less elongate, has the median line deeply marked but on posterior half only, and has the posterior angles rather less acute and produced. The punctures in the striæ of the elytra are also larger.

467.—MONOCREPIDIUS ATRATUS. n. sp.

Length 5 lines.

Black, subopaque, densely and finely punctate and clothed with very short fulvous pile. Head with a small longitudinal ridge on the vertex. Thorax subconvex, and very little longer than the width, with the median line slightly marked near the base only, and the posterior angles rather short, acute and strongly bicarinate. Scutellum as well as base of thorax clothed with ashen pile. Elytra about two and a half times longer than the width, rounded at the apex, and strongly striato-punctate. Antennæ and legs reddish brown, silky. Abdomen covered with a short sericeous fulvous pile.

468.—MONOCREPIDIUS MINOR. n. sp.

Length 4 lines.

Black, subnitid, punctate and clothed with long ashen pile. Head with a fine longitudinal ridge on the vertex. Thorax not longer than the width, with the median line indistinctly marked and the posterior angles large, acute, and bicarinated. Scutellum oblong, reddish. Elytra nearly three times longer than the width, striato-punctate, and narrowed and rounded at the apex. Legs yellow.

469.—MONOCREPIDIUS SUBMARMORATUS.   n. sp.

Length 4 lines.

Subelongate, black, slightly nitid, punctate, and clothed with ashen pile. Head reddish in front. Thorax longer than the width, with the median line very lightly marked, and the posterior angles strong, acute, bicarinated, and of a reddish colour. Scutellum elongate, red. Elytra three times longer than the width, striato-punctate, and rounded at the apex, with small patches, and rather indistinct fasciæ of whitish pile interspersed over their entire surface. Abdomen and legs of a brownish red.

470.—MONECREPIDIUS FULVIPENNIS.   n. sp.

Length 4½ lines.

Subnitid, finely punctate, and clothed with long fulvous pubescence. Head rather convex, black behind and red in front. Thorax very little longer than the width, not marked or obsoletely so on the median line, and of a brownish colour in front and red behind, with the posterior angles broad, acute, and strongly bicarinated. Elytra three times longer than the width, of a pale red colour, striato-punctate, and rounded at the apex. Body beneath and legs brownish red.

471.—MONOCREPIDIUS NEBULOSUS.   n. sp.

Length 3½ lines.

Subelongate, dark red, subopaque, densely punctate, and slightly pilose. Head subconvex. Thorax much longer than the width, without median line, and with the posterior angles moderately long, the appearance of having the apex broken off, and the inner carination very small. Elytra three times longer than the width, strongly striato-punctate, rounded at but not narrowed towards the apex, and somewhat cloudily marked with a broad sutural vitta and two broad fasciæ of a dark brown colour. Abdomen pale red. Legs yellow.

472.—MONOCREPIDIUS SUBFLAVUS. n sp.

Length $2\frac{1}{2}$ lines.

Pale yellow, subopaque, minutely punctate and pilose. Head dark brown, subconvex. Thorax much longer than the width, and brown at the apex—the brown colour sometimes extending in a point almost to the base in the middle and on the sides—with the posterior angles rather short, subacute, and bicarinated, the inner carina very minute. Scutellum nearly round. Elytra not three times longer than the width, striato-punctate, and rounded at the apex, with an elongate lateral spot behind the humeral angles, an oblong patch on the suture of the base, and a narrow zig-zag fascia behind the middle, of a dark brown. Body beneath brownish yellow.

473.—MONOCREPIDIUS SUBMACULATUS. n. sp.

Length $2\frac{1}{2}$ lines.

Brown, subopaque, finely and densely punctate, and covered with a short yellowish pile. Head subconvex. Thorax longer than the width, and pale coloured at the base, with the median line obsolete, and the posterior angles acute, unicarinate, and slightly curved upwards. Scutellum reddish. Elytra a little longer than twice the width, striato-punctate, and rounded at the apex, with some indistinct dull red patches over their surface. Legs pale yellow.

474.—MONOCREPIDIUS ALBIDUS. n. sp.

Length 4 lines.

Pale reddish brown, opaque, minutely punctate, and entirely covered with a very dense short whitish pubescence. Head nearly flat. Thorax longer than the width, without median line, and with the posterior angles rather short, acute, and strongly carinated. Elytra longer than twice the width, striato-punctate, and rounded at the apex. The parts of the mouth yellow.

475.—MONOCRPEIDIUS SUBGEMINATUS. n. sp.

Length $3\frac{1}{2}$ lines.

Reddish brown, subopaque, densely punctate and clothed with

Q

ashen pubescence. Head depressed in the middle. Thorax longer than the width, without median line, and with the posterior angles subacute and strongly bicarinated. Scutellum reddish, subolongate. Elytra nearly three times longer than the width, narrowed and rounded at the apex, and striato punctate—the alternate interstices being evidently broader. Body beneath dark brown, subnitid and finely punctate. Legs pale brown.

476.—Monocrepidius Canpezei. n. sp.

Length 4 lines.

Black, subopaque, densely punctate and clothed with a pale fulvous pubescence. Head subconvex. Thorax a little longer than the width, reddish at the base, without median line and with the posterior angles subacute and unicarinate. Elytra three times longer than the width, striato-punctate, and rounded at the apex. Legs pale yellow.

477.—Monocrepidius elongatulus. n. sp.

Length 3 lines.

Black, slightly nitid, punctate, and clothed with a short pale fulvous pubescence. Head minutely carinated on the top. Thorax much longer than the width, reddish at the base, without median line, and with the posterior angles subacute and bicarinate. Scutellum reddish, nearly round. Elytra three times longer than the width, striato-punctate, and rounded at the apex. Body beneath reddish brown, nitid, finely punctate, and clothed with a thin silky pubescence. Legs pale yellow.

478.— Monocrepidius castaneipennis. n. sp.

Length 5¼ lines.

Brown, punctate, subnitid, and clothed with a long thin yellowish pubescence. Thorax not longer than the width, without median line, and with the posterior angles rather short, subacute, and bicarinate. Scutellum subpentagonal. Elytra of a dark chesnut colour, twice as long as the width, narrowed and rounded at the apex, and striato-punctate, the punctures large and oblong. Abdomen reddish. Legs pale red.

479.—MEGAPENTHES AUTOMOLUS, Cand. *Mem. Soc. Roy. Liege Vol.* 14, *page* 495.

480.—ELASTRUS FLAVIPES. n. sp.

Length 4 lines.

Dark brown, subnitid, punctate, and sparingly pubescent. Head subconvex. Thorax longer than the width, with the median line deeply marked excepting near the apex, and with the posterior angles acute, carinated, curved outwards, and of a reddish colour. Scutellum dark red, nearly round. Elytra about three times longer than the width, rounded at the apex, rather coarsely punctate, and striato-punctate. Body beneath piceous brown, nitid, minutely punctate, and finely pubescent. Legs yellow.

481.—ELATER MASTERSII. n. sp.

Length 2¼ lines.

Blackish brown, subopaque, finely punctate, and clothed with short yellowish red pubescence. Head and thorax convex; the latter short, without median line, with the posterior angles short and obtuse, and with a sharp longitudinal impression on the base between the posterior angles and the middle. Scutellum large, nearly round. Elytra little longer than twice the width, rounded at the apex, and strongly striato-punctate. Antennæ palpi and legs pale red.

482.—CRYPTOHYPNUS VARIEGATUS. n. sp.

Length 1¾ lines.

Head and thorax subconvex, punctate, and densely covered with short black scales interspersed with some transverse golden scales. In form the thorax is little longer than the width, rounded on the sides, slightly narrowed behind, with the median line marked by a very minute smooth raised line, the posterior angles short and subacute, and a short fine ridge extending from the base upwards near and parallel to the posterior angles. The basal portion is mostly covered with silvery scales. Elytra rounded at the humeral angles, slightly widened towards the middle, narrowed and rounded at the apex, scarcely longer than twice the width, striate, and densely covered with scales, those near the scutellum being of a golden tinge, and on the sides and apex of a silvery white, while a broad fascia behind the middle,—which extends upwards in a narrow vitta along the suture and spreads outwards on each side into a half-circle,—and two sub-apical spots are formed of black scales. The antennæ have the first joint red, the remainder brown. The legs are pale red. The tarsi brown.

483.—CARDIOPHORUS MASTERSII. n. sp.

Length 2 lines.

Entirely testaceous yellow, opaque, finely punctate, and clothed with a long pale yellow pubescence. Thorax not longer than the width, and a little wider behind than in front, with the posterior angles short, subacute, and not carinated. Scutellum elongate-cordiform. Elytra paler than the thorax, very densely pubescent, not longer than twice the width, striato-punctate, and narrowed and rounded at the apex.

484.—CORYMBITES RUFIPENNIS. n. sp.

Length 4 lines.

Nitid, punctate, and black, with the basal articles of the antennæ, the palpi, the sides and under surface of the prothorax, the elytra, and the legs, red. Head with a curved transverse depression on the forehead. Thorax not longer than the width, with the median line marked near the base, and with the posterior angles rather short, acute, and carinate. Elytra longer than twice the width, roughly punctate, and deeply striated. Apex of abdomen reddish.

485.—CORYMBITES NIGRINUS. n. sp.

Length 3½ lines.

Differs from the last in size, in having the elytra less narrowed towards the apex, and in being entirely black excepting the posterior angles and under side of thorax, the abdomen, and the legs which are reddish, and the antennæ which are brown. Probably the male of the last species.

486.—OPHIDIUS BREVICORNIS. n. sp.

Length 8 lines.

Brassy black, densely punctate, and subopaque. Head with the border depressed in front, and a broad golden yellow vitta on each side between the eyes. Thorax elongate, wider behind than in front, bituberculate at the base, and profoundly sulcated in the middle, with three broad golden yellow vittæ on the back, and with the posterior angles long, acute, and carinate. Scutellum subtriangular, rounded at the apex. Elytra subelongate, subacuminate at the apex, striate with the interstices subcostate, and of a testaceous yellow colour barred and spotted with dark brown. Under side of body black, subnitid, and minutely punctate. Legs reddish brown. Antennæ black, short, thick, and strongly serrate.

Though resembling very much the other species of *Ophidius*, this insect differs considerably from all of them and from the characters of the genus given by M. Candeze. The antennæ are peculiar, the scutellum is not at all globular, and the tarsi are not dilated.

487.—ANILICUS SEMIFLAVUS, Germ. *Zeitschr*, V., *page*
163.—Cand. *Mon. vol.* 4, *page* 329.
anticus Dej. *Cat.* 3rd *Ed., page* 106.

488.—LUDIUS ATRIPENNIS. n. sp.

Length 2¾ lines.

Elongate, narrow, subnitid, punctate, clothed with semi-decumbent hairs, and of a black colour with the thorax above and below, the three basal segments of the abdomen and the legs red. Thorax with the median line lightly marked, and the posterior angles strongly carinate. Elytra striato-punctate, the punctures large and oblong.

489.—ACRONIOPUS RUFIPENNIS. n. sp.

Length 2¼ lines.

Brown, subopaque, punctate, and clothed with short semi-erect hair. Head and thorax convex, the latter not longer than the width, without median line, with the posterior angles rather short, and with these and the anterior margin of a dark red colour. Scutellum black, nitid, sparingly punctate and of oval form. Elytra as broad as the thorax, nearly parallel-sided, rounded at the apex, striato-punctate and of a pale red colour. Legs red.

490,—ACRONIOPUS PUBESCENS. n. sp.

Length 2 lines.

Subelongate, black, subnitid, punctate, and rather densely clothed with a whitish pubescence. Thorax not longer than the width, without median line, and with the posterior angles acute. Elytra subelongate, subacuminate towards the apex, and roughly punctate but not distinctly striate. Antennæ and legs reddish.

491.—ASCESIS MASTERSII. n. sp.

Length 5 lines.

Brown, subopaque, punctate and thinly clothed with whitish pile. Head subconvex, and with the frontal border nearly complete as in *Monocrepidius*. Thorax not longer than the width, without median line, and with the posterior angles acute, moderately long, carinate and slightly directed outwards. Elytra three times longer than the width, parallel-sided, rounded at the apex, and striato-punctate. Under side of body nitid, silky pubescent and minutely punctate. Legs and antennæ reddish brown.

I am probably wrong in placing this insect in the genus *Ascesis*. The antennæ are shorter, thicker, and more strongly dentated than in the typical species. The legs are much more thick and short, and the head is more like that of a *Monocrepidius*, than of any of the group in which *Ascesis* is placed.

492.—DICTENIOPHORUS VITTICOLLIS. n. sp.

Length ♂ 4 lines.

Black, subnitid, minutely and densely punctate, and densely clothed with a very short fulvous pubescence. Thorax longer than the width, red on the sides, and hoary pubescent at the base, with the median line marked on the posterior half, and the posterior angles acute and carinate. Elytra pale red, subacuminate towards the apex, and striato-punctate. Antennæ shorter than half the body, and strongly pectinate.

493.—DICTENIOPHORUS APICALIS. n. sp.

Length 5 lines.

Black, subnitid, punctate and clothed with very short pile. Thorax red with basal margin black, and scarcely longer than the width, with the median line deeply marked on the posterior half, and the posterior angles acute, carinate, curved slightly outwards, and of a black colour. Elytra subelongate, subacuminate at the apex, striato-punctate, and of a pale red colour on the basal two-thirds.

494.—DICTENIOPHORUS VITTATUS. n. sp.

Length ♂ 3 lines.

Elongate, narrow, black, subnitid, punctate, and moderately pubescent. Thorax much longer than the width, with the median line only marked at the very base, and the posterior angles obtuse and carinate. Elytra striato-punctate with a broad lurid yellow vitta along the whole length of each elytron. Legs reddish brown.

HEMIOPSIDA. n. gen.

Last joint of maxillary palpi small, subovoid. Head rather prominent, vertical, excavated, and broadly rounded in front. Eyes round, entirely disengaged from the thorax. Antennæ moderately long, first joint thick, second very small, third also small but larger than the second, 4 to 10 long and dentated, 11 very long and filiform. Thorax short, broadest at the base. Scutellum oblong, subtruncate. Legs thick, tarsi entire—1st joint longest, the rest gradually decreasing to the fourth. Body rather robust. Prosternum convex and without mentonniere.

495.—HEMIOPSIDA MASTERSII. n. sp.

Length 4¼ lines.

Convex, black, subopaque, punctate, and clothed with short

golden yellow pile. Antennæ with the exception of the basal joint of a brownish red. Head deeply excavated in front between the antennæ. Thorax shorter than the width, with the median line slightly marked by a smooth linear space, a small fovea in the middle between the median line and the sides, and the posterior angles short and obtuse. Elytra dark red, coarsely punctate, and striate with the interstices elevated. Tarsi and extremities of the tibiæ brownish red.

MALACODERMIDÆ.

496.—METRIORRHYNCHUS RHIPIDIUS, MacL.   *App.*
King's Surv., page 442.
    ♀ *septemcavus,* MacL.   *App. King's Surv.,*
    page 443.

497.—METRIORRHYNCHUS FEMORALIS.   n. sp.

Length ♂ 5 lines, ♀ 7 lines.

Antennæ black, dentate from the third article, and alike in both sexes. Head small, black, nitid, and deeply impressed. Thorax red with black centre, and divided as in the last species into 7 hollows, with the posterior angles pointed backwards more acutely than usual in the genus. Scutellum black, nearly square, emarginate at the apex. Elytra a little broader than the thorax at the base, and of an orange red colour tipped in the female with dark blue, with four fine costæ on each, and with the intervals filled with shallow square punctures disposed in double rows. Body beneath black. Legs black, excepting the coxæ and basal two-thirds of the thighs which are red.

498.—METRIORRHYNCHUS NIGRIPES.   n. sp.

Length 5 lines.

This species differs from the last in having the head only lightly impressed on the median line, and not nitid, in having the thorax more rounded at the posterior angles and black only on the basal portion of the middle, in having the elytra of a darker red and more deep punctation, and in having the legs entirely black.

499.—METEIORRHYNCHUS MARGINICOLLIS. n. sp.

Length 4 lines.

Black. Antennæ strongly dentate. Head transversely and deeply impressed between the eyes. Thorax 7-hollowed, and subtruncate at the base, with the lateral margins yellowish red. Scutellum deeply emarginate at the apex. Elytra yellowish red four-costate or counting the lateral border and the suture six-costate, with the intervals rather confusedly punctate in double rows, and towards the apex appearing to consist of single rows of transverse punctures. Legs and under surface of body entirely black.

500.—CALOCHROMUS GUERINII. n. sp.

Length 4 lines.

Bluish black, nitid. Thorax little broader than the head, subtransverse, and quadrangular, with the angles rounded, the median line deoply impressed and foveate near the base, and the sides broadly bifoveate and of an orange colour. Scutellum sub-truncate. Elytra orange red with the apex blue, and closely subcostate, the alternate costæ slightly larger than the others.

I have named this species after the founder of the genus, who nearly forty years ago formed it for the reception of an insect of New Guinea. It is not by any means an uncommon Australian form.

501.—LUCIOLA FLAVICOLLIS. n. sp.

Length 3 lines.

Head nearly covered by the thorax, black, subopaque, sub-depressed, and canaliculated in the middle, with the eyes in the male very large. Thorax luteous yellow, transverse, subquadrangular, largely lobed in the middle at the apex, slightly so at the base, and punctate, with the median line distinct, a transverse depression near the apex and base, and a shallow elongated fovea near the sides. Scutellum yellow, subtriangular, rounded at the apex. Elytra broader than the thorax, punctate, and of a dark brown colour, with the base, the suture as far as the middle, and the sides almost to the apex, of a luteous yellow. Body beneath black, excepting the prothorax, the coxæ, and the basal portion of the thighs which are red, and the penultimate segment of the abdomen, which is of a waxy white.

502.—TELEPHORUS FLAVIPENNIS. n. sp.

Length 4½ lines.

Black, subnitid. Head with a yellow transverse spot at the insertion of the antennæ, the second joint of these very short. Thorax square, and margined all round, with the angles and the apex rounded and the base minutely emarginate, the whole bordered with yellow. Elytra yellow, dehiscent, subacuminate, and confusedly punctate, with two very fine costæ on each, the inner one short, the outer extending nearly to the apex. The coxæ and the apex of the abdominal segments are yellow.

503.—TELEPHORUS RUFICOLLIS. n. sp.

Length 4½ lines.

Head large, black, with a red spot on the vertex, and all in front of the insertion of the antennæ yellow. The antennæ are with the exception of the under side of the basal joint, brown. Thorax subtransverse, not narrowed behind, rounded at the angles, emarginate in the middle of the base, and of a dark red colour becoming yellow on the borders. Elytra bluish green, subnitid, granulose. Pro-meso, and meta-thorax, and apex of all the abdominal segments, yellow, coxæ and basal half of thighs, red.

504.—TELEPHORUS MASTERSII. n. sp.

Length 4½ lines.

Broad, pale red, subnitid. Antennæ robust. Thorax transverse, broadest at the base and deeply impressed at the posterior angles. Scutellum subtriangular, rounded at the apex. Elytra confusedly punctate, obsoletely striate and of a yellow colour with the base and apex black. Antennæ, palpi and legs excepting the coxæ, black.

505.—ICHTHYURUS DEPRESSICOLLIS. n. sp.

Length 3 lines.

Black, subnitid. Antennæ brown with the last three joints

pale 'red. Thorax rather longer than the width, rounded at the apex, base, and angles, transversely depressed in the middle, and of a yellow colour with a broad black band in the middle. Scutellum yellow. Elytra short, dehiscent, subacuminate, and confusedly punctate, with an elongate yellow patch towards the base, and a broad fascia of the same colour near the apex. Prothorax beneath, coxæ, and apical edge of the abdominal segments yellow. Tibiæ reddish brown.

506.—LAIUS BELLULUS, Guer. *Voy. Coq.*, *page* 78, —Germ. *Linn. Ent.* III., *page* 182.

507.—LAIUS MASTERSII. n. sp.

Length 2 lines.

Cyaneous, nitid, punctate, hairy. Thorax red, transverse, deeply impressed along the base. Elytra with a broad golden fascia in the middle.

This species differs from *L. bellulus* chiefly in being of a deeper blue, in being less strongly punctate, more hirsute, and in having no apical spot on the elytra, while the fascia is broader, straighter, and of a more yellow hue.

508.—MALACHIUS LURIDICOLLIS. n. sp.

Length 2¼ lines.

Oblong, flat, black, subnitid, indistinctly punctate, and clothed with short erect fulvous pubescence. Head nitid, with the muzzle and three first joints of the antennæ pale red. Thorax transverse, nearly truncate at the apex, rounded at the anterior and subacute at the posterior angles, broadly rounded at the base, and of a lurid red colour, with the margins elevated and with two small deep round foveæ at the base. Elytra as long as the abdomen, broadest towards the apex, and of a blackish brown colour tinged with lurid brown at the base. Legs yellow, the fourth joint of the tarsi lobed beneath.

509.—CARPHURUS CYANEIPENNIS. n. sp.

Length 4 lines.

Narrow, pale red, nitid and hairy. Head foveolate in front,

with the first joint of the antennæ large and emarginate on the upper surface. Thorax of the width of the head, truncate at the apex, and gradually rounded from the anterior angles along the sides, posterior angles, and base, and with a broad transverse depression near the apex and base. Elytra half the length of the abdomen, separately rounded at the apex, irregularly punctate and of a blue colour. Abdomen above and below with the last three segments red, the others black bordered with red. Legs and lateral margin of abdominal segments clothed with long hair.

510.—CARPHURUS ELONGATUS. n. sp.

Length 4 lines.

Narrow, black, nitid, and hairy. Antennæ short, submoniliform, with the two first joints red. Head broadly bicanaliculate in front, and of a dark red colour. Thorax rather narrower than the head, longer than the width, rounded in front and behind, margined all round and of a dark red colour, with a black patch on each side, and a broad transverse depression near the apex and base. Scutellum small, transverse. Elytra very short, rather broader than the thorax, lightly punctate, and subtruncate, with the basal half dark red. All the segments of the abdomen bordered with red. Tibiæ reddish brown.

511.—CARPHURUS APICALIS. n. sp.

Length 2 lines.

Elongate, subnitid, moderately hairy and of a red colour, with the six apical joints of the antennæ, the back of the head, the apical two-thirds of the elytra, and the two apical segments of the abdomen, black. Thorax narrower than the head, longer than the width, broadest and subtruncate at the apex, constricted and transversely impressed near the base, and moderately rounded behind. Elytra short, considerably broader than the thorax, and broadest at the apex which is subtruncate.

512.—CARPHURUS AZUREIPENNIS. n. sp.

Length 3¼ lines.

Elongate, nitid, clothed with long black hair. Head red, exserted, narrowed behind and deeply bicanaliculate between the eyes, with the first two joints of the antennæ red, the rest black and subdentate. Thorax red, about the width of the head and truncate at the apex, longer than broad, slightly narrowed above the posterior angles, and slightly rounded and margined at the base, with a transverse depression immediately in front of it. Scutellum dull red, transverse. Elytra short, deep blue, punctate and truncate. Abdomen with the two first segments red, the remainder black. Legs entirely black.

513.—CARPHURUS PALLIDIPENNIS. n. sp.

Length 2½ lines.

Elongate, nitid, thinly clothed with long blackish hair. Head rather large, exserted, slightly narrowed behind the eyes, lightly bicanaliculate in front, thinly punctate and of a red colour, with the labrum, and the antennæ excepting the two first joints, black. Thorax red with black sides, narrower than the head, longer than the width, truncate in front, rounded behind, transversely impressed near the base and margined on the sides and base. Scutellum transverse, broadly rounded at the apex. Elytra short, pale red, punctate, and obliquely truncate. Abdomen black, with the two terminal segments piceous. Legs black, with the tibiæ more or less red.

BALANOPHORUS. n. gen.

Maxillary palpi fusiform, obtuse. Labrum transverse, rounded in front. Head broad. Eyes large and prominent. Antennæ rather short, first joint much larger than the second, the third dentate, the remainder pectinate. Elytra much shorter than the abdomen. Tarsi short, first joint large, the three following very small; in the anterior tarsi the second takes its rise from the middle of the base of the first. The visicles of the thorax and abdomen large and exserted.

514.—BALANOPHORUS MASTERSII. n. sp.

Length 4 lines.

Nitid, thinly punctate, very hairy. Head black and impressed with a transverse curve on the forehead, with the anterior

part and three first joints of the antennæ red, and with the rest
of the antennæ and the tooth of the third joint black. Thorax
red, nearly truncate in front, slightly narrowed behind, rounded
and with a recurved margin at the base, and with the thoracic
vesicles flat and extending along the sides nearly to the apex.
Scutellum black. Elytra about half the length of the abdomen,
broader than the thorax, lightly punctate, subtruncate and of a
chalybeate blue, with the base red. The abdomen has the basal
joints red, the apical black, but the lateral vesicles on all are red.
The coxæ of the anterior legs are red.

Two other species of this family are in the collection, but, as
they are single specimens and gummed down on cardboard, I
cannot undertake to describe them. One looks like a *Laius*
though with eleven joints to the antennæ, the other resembles a
small *Malachius* with short elytra and very long hind legs. They
will both probably be found to be new genera. Indeed I believe
that the *Malacodermidæ* of Australia, though not very numerous,
will, when properly investigated, exhibit a number of new
and very curious genera, particularly among the subfamily
Malachidæ.

CLERIDÆ.

515.—CYLIDRUS CENTRALIS, Pasc. *Journ. of Ent.* I.,
page 44.

516.—CYLIDRUS BASALIS. n. sp.

Length 4 lines.

Black, nitid, hairy. Head coarsely punctate with the palpi
and basal joints of the antennæ, red. Thorax lightly punctate.
Elytra about half the length of the body, rounded at the apex,
very slightly punctate and with the basal half of a dark red
colour. Legs and metathorax yellow.

517.—OPILUS CONGRUUS, Newm. *The Entomol.* 1842,
page 365.
var. *femoralis*, White. *Cler.,* IV., page 55.

518.—OPILUS INCERTUS.   n. sp..

Length 2¼ lines.

Brownish black, subnitid, coarsely punctate and hairy.   Eyes
small.   Basal joints of antennæ red.   Thorax broadly and longitudinally impressed in the middle and without transverse
impression.   Elytra rounded at the apex and slightly wider there
than at the base, strongly striato-punctate on the anterior half,
and almost smooth towards the apex, with a pale yellow fascia
just behind the middle, and not reaching the suture.

519.—NATALIS CRIBRICOLLIS, Spin. *Mon.* 1, *page* 203,
    t. 16. *f.* 4.

520.—NATALIS PORCATA, Fabr.   *Mant. Ins.* I., *page* 127.
    —Klug. *Mon.*, *page* 318.—Spin. *Mon.* 1,
    page 201, *t.* 16. *f.* 2.
    heros, Sturm. *Cat.* 1843, *page* 82.

521.—NATALIS MASTERSII.   n. sp.

Length 8 lines.

This species differs from *N. porcata* in being of a darker
colour, and more elongate form.   The thorax differs in being
much longer than the width, in being constricted in the middle,
and in having a prominent tubercle or bulge between that and
the base.   The elytra differ in being less hairy, in being more
regularly punctate, and in having the alternate interstices only
elevated.

522.—STIGMATIUM GILBERTI, White.   *Mus. Cat.*
    Clerid. 1849., *page* 53.

523.—STIGMATIUM MASTERSII.   n. sp.

Length 5½ lines.

Black, subnitid, moderately hairy, and densely clothed in
patches with a silky yellow pubescence.   Head in front clothed
with white hair, with the palpi yellow, and the antennæ slender.—
the basal joints reddish and the third joint the longest.   Thorax
entirely covered with yellow hair.   Elytra very coarsely punctate

in double distant rows at the base, and very minutely punctate towards the apex, with the basal portion marked with small patches of white and yellow pubescence, with an intervening smooth patch, with a broad post-median fascia of golden yellow pile changing in some lights to an olive green, and with the apex clothed in the same way. Basal half of thighs yellow. Tibiæ and tarsi reddish brown.

524.—STIGMATIUM LÆVIUS. n. sp.

Length 4 lines.

This species is of a much flatter and less robust form than the last, and seems from the description to approach very nearly to an insect from Prince of Wales Island described by Professor Westwood under the name of *Omadius olivaceus*.

The upper and under surface is of a subnitid piceous brown with the exception of the basal joints of the antennæ, the palpi, the basal portions of the thighs, the tibiæ and the tarsi, which are reddish. The head and thorax are covered with yellowish hair interspersed with strong setæ. Elytra flat, broadest about the middle, substriate, thinly punctate, and setose in rows, with the basal half covered with pale yellow pubescence, and with an irregular fascia behind the middle and the apex similarly marked. Hind legs long.

525.—STIGMATIUM VENTRALE. n. sp.

Length 3 lines.

Black, densely punctate, hairy. Head clothed in front with white hairs and on the top with a golden yellow pubescence. Thorax with the sides flavo-pubescent. Elytra striato-punctate, setose,—the setæ on the sides white—and granulose with a large chocolate coloured patch in the middle which extends near the sides to the humeral angles and is bordered by whitish hairs, and an irregular semi-circular patch and the apex cinereo-pubescent. The meta-thorax and abdomen are red. The legs reddish brown.

526.—OMADIUS PRASINUS, Westw. *Proc. Zool. Soc.*, 1852, *page* 53, *t.* 26, *f.* 2.

527.—THANASIMUS EXIMIUS, White. *Cat. Mus. Cleridæ,*
1849, *page* 63.—Westw., *Proc. Zool. Soc.,*
1862, *page* 54.

528.—THANASIMUS SCULPTUS. n. sp.

Length 3½ lines.

Head and thorax brassy black, subnitid, finely punctate, and
clothed with white hair. Elytra much broader than the thorax,
closely marked except at the apex with rows of quadrangular
transverse excavated punctures, and of a beautiful coppery red
colour, with a broad bluish purple fascia in the middle. Body
beneath dark blue. Legs purplish.

529.—CLERUS SEPULCHRALIS, Westw. *Proc. Zool.
Soc.,* 1852, *page* 52, *t.* 25, *f.* 9.

530.—CLERUS NOVEMGUTTATUS. Westw. *Proc. Zool.
Soc.,* 1852, *page* 49.

531.—CLERUS MASTERSII. n. sp.

Length 3 lines.

Dark blue, nitid, punctate and hairy. Thorax with a broad
rugose depression in the middle of the median line. Elytra
broader than the thorax, profoundly punctate except towards the
apex, and of a bronzy red colour, with a brighter spot near the
apex, and with a fascia above the middle, the suture from that to
the base, and the base itself for half its width, yellow. Antennæ
palpi and tarsi pale brown.

532.—CLERUS APICALIS. n. sp

Length 2¾ lines.

Black, subnitid, coarsely punctate and hairy. Thorax impressed in the middle. Elytra very profoundly punctate almost
to the apex, with a narrow white fascia behind the middle, and a
large luteous spot at the apex. Body beneath and legs blue.
Antennæ and palpi reddish.

K

533.—Aulicus instabilis, Newm. *The Entomol.*, *page*
15—Klug. *Mon.*, *page* 341.—Spin. *Mon.* 1,
page 331, *t.* 28, *f.* 1.
var. *castanipes and tibialis*, Westw. *Whites*,
Cleridæ, IV., *page* 60.

534.—Aulicus rufipes. n. sp.

Length 2½ lines.

Shorter and broader than *A. instabilis.* Head and thorax golden green, the latter very hairy and deeply impressed transversely and in the middle of the median line. Elytra green and punctate, the punctures large towards the base, but nowhere very deep or regular. Legs and the parts of the mouth entirely pale red.

535.—Aulicus foveicollis. n. sp.

Length 2¾ lines.

Of a more elongate form than the last, entirely of a dark blue with the legs reddish brown, very strongly punctate on the elytra and with an almost triangular fovea in the centre of the thorax.

536.—Aulicus viridissimus, Pasc. *Journ. of Ent.*
1, *page* 47.

The specimen before me is smaller than that Mr. Pascoe describes, and has the anterior tarsi of a reddish colour, but in other respects they seem identical.

537.—Tarsosternus pulcher. n. sp.

Length 3 lines.

Elongate, red, nitid, thinly punctate and hairy. Head large, and convex, with the eyes elongate. Thorax not broader than the head, longer than the width, and narrowed at the base. Elytra strongly punctate on the basal half, with a very broad blue fascia about the middle not reaching the suture and with, immediately behind it, a narrow smooth raised white fascia. Under side of body and legs reddish brown. Antennæ, basal joints excepted, dark brown.

This insect has very much the appearance of a *Tillus.*

538.—Tarsosternus Mastersii. n. sp.

Length $2\frac{1}{2}$ lines.

Elongate, densely punctate, and moderately hairy. Head and thorax bluish green, coarsely punctate, the former sub-convex, with the eyes large and coarsely granulate, the latter rather narrower than the head, much longer than the width, and narrowed at the base. Elytra very little broader than the thorax, three times longer than the width, and densely foveate on the basal half, with the base of a brilliant purple red, with a broad dark green fascia behind, with a narrow oblique white fascia next, with the remainder of the elytra of a burnished purple red, and with an apical vitta consisting of white hair. Abdomen beneath black. Legs violet red.

539.—Trogodendron fasciculatum, Schreib. *Trans. Lim. Soc.*, 1802, VI., *page* 195, *t.* 20, *f.* 6. —Klug. *Mon.*, *page* 326.—Spin. *Mon.* 1, *page* 212, *t.* 18, *f.* 1.
var. *honestum*, Newm. and Lacord.

540.—Scrobiger splendidus, Newm. *The Entomol.*, 1840, *page* 15.
Reichei Spin. *Mon.* 1, *page* 232, *t.* 14, *f.* 1.

541.—Scrobiger albocinctus, Pasc. *Journ. of Ent.* 1, 1860, *page* 46.

542.—Zenithicola obesa, White. *Stokes voy. app.*, *t.* 1, *f.* 9, var. *obesula* White. *Clerid. Cat. Mus.*, *page* 26.

543.—Eleale lepida, Pasc. *Journ. of Ent.* 1, 1860, *page* 45.

544—Eleale fasciata. n. sp.

Length 5 lines.

Black, subnitid, very densely and coarsely punctate. Antennæ pale red. Scutellum covered with white hair. Elytra with a broad orange fascia above the middle and a minute apical spot of white hair. Under side of body thickly clothed with white hair. Legs cyaneous.

545.—ELEALE APICALIS. n. sp.

Length 5 lines.

Elongate, nitid, densely punctate. Head and thorax of a bronze hue, the latter flat on the back, rugosely and transversely punctate, longer than the width, constricted and transversely impressed near the apex and narrowed at the base. Elytra coarsely and irregularly punctate with three slightly raised lines on each, and of an orange yellow colour with the apex black. Body beneath dark blue or purple with metallic reflection, and spotted with tufts of white hair. Antennæ yellow.

546.—ELEALE ELONGATULA. n. sp.

Length 5 lines.

Of more elongate form than *E. aspera* Newm. and densely and coarsely punctate. Head and thorax greenish blue, the latter purple on the sides, twice longer than the width and cylindrical. Elytra about four times longer than the width, and of a golden green with a broad purple lateral margin. Legs of a purple red. Body beneath spotted with tufts of pale pubescence. Antennæ reddish brown, the apical joint deeply emarginate at the inner angle.

547.—ELEALE VIRIDICOLLIS. n. sp.

Length 3½ lines.

This species differs from the last in being much smaller, in having the head and thorax golden green, and the latter less elongate and less cylindrical, in having the elytra entirely of a dull purplish green, in having the under portion of the body thickly clothed with white hair, but without the white spots of the other, and in having the last joint of the antennæ more largely but less deeply emarginated.

548.—LEMIDIA HILARIS, Newm. *Zoologist*, 1843, *page* 119.
corallipennis, Westw. *Proc. Zool. Soc.*, 1852, *page* 47.

549.—ALLELIDEA CTENOSTOMOIDES, Waterh. *Trans. Ent. Soc.* II., *page* 194, *t.* 17, *j.* 1.

550.—TENERUS RUFICOLLIS. n. sp.

Length 3 lines.

Nitid, very finely punctate, and moderately hairy. Head black, with a dull red spot in the middle and at the insertion of the antennæ. Thorax red, subquadrate, transversely impressed before the middle, not broader than the head. Elytra shorter than the body and blue, with a broad ill defined fascia behind the middle, and the apex, of a paler hue. Under side of head and prothorax pale red. Legs, meso- and meta-thorax, and three apical segments of abdomen, dark blue. Basal segments of abdomen deep red. Antennæ black, and subpectinate from the fourth joint.

551.—PYLUS PALLIPES. n. sp.

Length 3¼ lines.

Much smaller and of a less elongate form than *P. fatuus* Newm. Pale chesnut, subnitid, coarsely punctate. Head brown, finely punctate, with the eyes large, granulose, and minutely emarginate in front. Thorax subquadrate, slightly constricted behind the anterior angles, suddenly and largely bulged out in the middle, and rectangular and truncate at base, with the punctures thin and one or two shallow impressions on the disk. Elytra much broader than the thorax, and punctate in regular rows,—the punctures large. Legs pale reddish yellow.

552.—NECROBIA VIOLACEA, Linn. *Syst. Nat. Ent.* 10, *page* 356.—Klug. *Mon.*, *page* 349.—*Spin. Mon.* 11, *page* 105. Syn. *angustata.* Falderm. *Chalybea.* Sturm. *Cyanella* Anders., *errans.* Welsh. *quadra.* Marsh.

PTINIDÆ.

553.—Ptinus albomaculatus. n. sp.

Length 1¼ lines.

Oval, black, subnitid, punctate, and clothed with erect black hairs. Antennæ and legs variegated with white pubescence. Elytra striato-punctate, the interstices flat and broad, with a round spot near the humeral angles and a transverse one behind the middle of a yellowish white colour.

BOSTRYCHIDÆ.

554.—Rhizopertha elongatula. n. sp.

Length 2 lines.

Elongate, black, subnitid, thinly punctate. Head with a round fovea in the centre of the forehead, and with the antennæ and palpi of a piceous red. Thorax longer than the width, rough, toothed, and somewhat retuse in front and smooth behind. Elytra three times longer than the width, punctate, and deeply emarginate on the external angle of the apex, with a smaller emargination at the apex. Legs red.

555.—Rhizopertha gibbicollis. n. sp.

Length 1¾ lines.

Of a short oblong form, black, subnitid, and coarsely punctate. Thorax broader than the length, very convex, very rough, dentated in front, and smooth behind. Elytra piceous red on the basal half, and flatly sloped away from near the middle to the apex, the flat surface margined all round except near the suture on the upper part. Under side of abdomen clothed with a fine white pubescence. Legs and antennæ piceous red.

556.—Bostrychus bispinosus. n. sp.

Length 1½ lines.

Oblong, piceous black, subnitid, and punctate. Thorax not longer than the width, rough in front and smooth behind. Elytra rather roughly punctate, and cut very flatly and steeply behind, with a large subacute spine rising from the suture on each elytron near the apex, and extending in a direction backwards and outwards. Legs and antennæ piceous red.

557.—BOSTRYCHUS CYLINDRICUS. n. sp.

Length 4 lines.

Elongate, piceous black, nitid and punctate. Head densely and minutely punctate. Thorax slightly longer than the width, rough and dentated in front,—the anterior tooth hooked—and smooth behind. Elytra coarsely punctate towards the apex, and having on the retuse portion six sharp projections, three on each elytron, one near the suture, another about the middle and the third on the side near the apex, which is obliquely truncate and separately rounded. Legs, antennæ, and underside of body, piceous.

558.—BOSTRYCHUS JESUITUS, Fabr. *Ent. Syst.* 1, 2, *page* 361.—Boisd. *Voy. Astrol.* II., *page* 461.

TENEBRIONIDÆ.

559.—OPATRUM MASTERSII. n. sp.

Length 4 lines.

Ovate, subdepressed, brown, opaque, and covered with short yellow setiform scales. Head transversely impressed in front, and prominently angled in front of the eyes. Thorax much broader than the length, broader behind than in front, rounded on the sides, broadly rounded at the base in the middle, and emarginate near the angles, with the anterior angles prominent, the posterior very acute and slightly pointed backwards and outwards, the median line visible, and a broad subconcave lateral border. Scutellum transverse, rounded behind. Elytra of the width of, or slightly wider than, the thorax, bisinuate at the base, and marked with broad smooth striæ, with the interstices broad, subconvex, and covered with about three irregular rows of yellow setiform scales.

560.—APATELUS SQUAMOSUS. n. sp.

Length 3½ lines.

Oblong, black, opaque, punctate, squamose. Head canaliculate on the suture of the epistome, and very slightly emarginate in front. Thorax subquadrate, slightly rounded on the sides, and truncate at the base, with the anterior angles moderately prominent and the posterior subacute and a little recurved. Elytra broader than the thorax, rounded at the humeral angles, and striato-punctate with the interstices subcostate. Body beneath piceous black, subnitid and punctate. Tarsi piceous and clothed with reddish hair.

561.—CESTRINUS SQUALIDUS. n. sp.

Length ♂ 5, ♀ 6 lines.

Black, opaque, roughly punctate, and very squamose. Thorax transverse with broad recurved lateral margin, much rounded on the sides, and truncate at the base, with the anterior angles prominent and subacute, and the posterior acute and pointed a little outwards. Elytra broader than the thorax, rounded at the humeral angles, convex, nearly perpendicular on the apical third, and marked with large elongate punctures disposed in irregular rows, and elevated alternate interstices covered with black nitid granules. Body beneath less squamose than above. Tarsi piceous.

562.—HYOCIS PALLIDA. n. sp.

Length 1½ lines.

Oval, subconvex, pale red, subnitid, and punctate—each puncture furnished with a very short decumbent yellow seta. Head slightly impressed transversely between the eyes. Thorax transverse, nearly truncate in front, broadly rounded at the base, and slightly rounded at the sides, with the median line distinct. Scutellum small, rounded. Elytra of a yellow colour, rather broader than the thorax at the base, becoming slightly broader towards the apex, and profoundly striato-punctate. Fore tibiæ large and flat.

563.—HYOCIS PUBESCENS. n. sp.

Length 1¼ lines.

This species differs from the last in its very distinct kind of pubescence, which is rather long, decumbent, and of a white colour. It differs also in having the head of a brassy black and without impressions, in having the thorax with the anterior angles advanced, and two longitudinal foveæ at the base, in having the scutellum triangular, and in having the elytra less nitid and marked with a few brown spots.

564.—MYCHESTES PASCOEI. n. sp.

Length 4 lines.

Black, opaque, tuberculose, granulose and densely covered with brownish yellow scales. Head foveolate in front, with the suture of the clypeus semicircular, the parts of the mouth nitid, and the labrum emarginate. Thorax subtransverse, emarginate in front, truncate behind, much bulged out and abruptly rounded on the sides and elevated on the disk into a large laterally compressed rounded tubercle which projects over the head and which extends itself backwards in a triangular form nearly to the base of the thorax where it terminates in an obtuse tubercle. Scutellum nearly round. Elytra convex, scarcely longer than the width at the base, of the same size as the base of the thorax and fitting closely to it, swelling out in the middle to the size of the thorax at its broadest part, subacuminate and perpendicular towards the apex, and rough on the surface, with large depressions and obtuse tubercles, the most elevated being a three headed tubercle near the base on each side and some distance from the suture.

565.—MYCHESTES MASTERSII. n. sp.

Length 4 lines.

Oblong-oval, black, opaque, rough, rugosely tuberculate, and densely covered with dark brown scales. Thorax truncate at the base, broadly marked on the median line, and advanced in front into a round rough projection constricted behind which entirely covers the head, and looks from above exactly like a head and neck. Scutellum small, rounded. Elytra truncate and of the same width as the thorax at the base, becoming a little wider towards the apex, and covered with obtuse tubercles.

566.—PLATYDEMA ARIES, Pasc. *Ann. Nat. Hist.*, 1869, *page* 280.

567.—PLATYDEMA LIMACELLA, Pasc. *Ann. Nat. Hist.*, 1869, *page* 280.

568.—PLATYDEMA PASCOEI. n. sp.

Length 1½ lines.

Black, subnitid, very finely punctate. Head transversely impressed in front of the eyes. Thorax slightly foveolate at the base half way between the median line and the posterior angles. Elytra finely and distinctly striato-punctate, with the base and apex of a piceous red. Legs reddish brown.

This species is more elongate and less convex than the preceding two species and is moreover quite unarmed on the head, unless the specimens before me are all females.

569.—PLATYDEMA LATICOLLE. n. sp.

Length 2 lines.

This insect differs considerably from the other species of Platydema in having a flatter and more oval form, longer antennæ and larger palpi ; it ought probably to form a new genus.

The whole insect is of a piceous hue, very nitid and very minutely punctate, excepting on the elytra which are distinctly striato-punctate. The thorax is broad, more truncate at the base than usual in the genus, and distinctly marked with longitudinal impressions, instead of foveolæ at the base.

570.—CEROPRIA PEREGRINA, Pasc. *Journ. of Ent.*, II., *page* 460.

571.—TYPHOBIA FULIGINEA, Pasc. *Ann. Nat. Hist.*, 1869, *page* 279.

572.—ACHTHOSUS LATICORNIS, Pasc.   *Ann. Nat. Hist.*,
1869, *page* 294.

573.—TOXICUM DISTINCTUM.   n. sp.

Length 5½ lines.

Black, subopaque, finely punctate, except on the elytra.
Head with the anterior horns slight, acute, and half a line long,
the posterior large, thick, obtuse, clothed with yellow hairs on
the apex and anterior edge, and projecting upwards, forwards,
and inwards at the apex.   Thorax subtransverse, parallel-sided,
very slightly lobed in front and behind, and marked near the
middle with two small foveæ.   Elytra a little broader than the
thorax, and strongly punctate in regular rows.   Tarsi piceous.
Club of antennæ of three joints.

574.—TOXICUM PARVICORNE.   n. sp.

Length 5 lines.

Black, opaque, densely punctate.   Head with the anterior
horns represented by small acute tubercles, and the posterior
rather short, broad and laterally compressed at the base, subacute
and with a small yellow tuft on the apex, and directed almost
straight upwards, and inwards at the apex.   Thorax subtransverse, parallel-sided, very slightly lobed in front and behind,
slightly prominent and rounded at the anterior angles, and acute
at the posterior.   Elytra a little broader than the thorax, and
marked with regular rows of small but deep punctures.   Under
side of body thinly punctate.   Tarsi piceous.   Club of antennæ
of three joints.

575.—PTEROHELÆUS ASELLUS, Pasc.   *Ann. Nat. Hist.*,
February, 1870.

576.—PTEROHELÆUS BREMEI.   n. sp.

Length 10 lines ; width 6½ lines.

Broadly ovate, black, subopaque.   Head large, subquadrate,
truncate in front, and rounded at the angles, with a broad shallow
canaliculation between the eyes.   Thorax transverse and

largely emarginate in front, with large flat margins, a little raised and thickened on the border towards the anterior angles, and with the posterior subacutely pointed backwards. Scutellum transversely and curvilinearly triangular and transversely impressed in the middle. Elytra not longer than the width, as broad as the thorax at the base, and rounded at the apex, with a broad smooth margin—broadest at the humeral angle and becoming narrower to the apex—raised on the border, and with the disk marked with eight subcostate elevations, the second from the suture the largest, the lateral ones resembling continuous rows of nodules, and the intervals rather obliterately punctate in double rows. Abdomen subnitid, and marked with longitudinal striolæ. Antennæ, palpi, and tarsi, piceous.

577.—PTEROHELÆUS ELONGATUS. n. sp.

Length 10 lines, width $4\frac{1}{2}$ lines.

Oblong-oval, black, subopaque. Head transverse, punctate, widened in front of the eyes, rounded at the anterior angles, and almost truncate in front, with a narrow recurved margin. Thorax with a broad lateral margin a little reflexed at the anterior angles, and with the posterior angles less pointed backwards than in the last described species. Scutellum triangular. Elytra nearly twice as long as the width, narrowly and equally margined, and marked with eight costiform elevations alternating with smaller ones, some of which are scarcely traceable, with the intervals strongly punctate. Under surface nitid, substriolate. Antennæ, palpi, and tarsi, piceous.

578.—PTEROHELÆUS PASCOEI. n. sp.

Length 9 lines ; width 6 lines.

Broadly ovate, black, opaque. Head scarcely enlarged before the eyes, broadly rounded in front, finely canaliculate between the eyes, and with a semi-circular line or suture extending across, and to the front of, the head before the eyes. Thorax with a broad flat margin, and a lightly marked median line. Scutellum curvilinearly triangular. Elytra broadly margined,—the margin of a dull reddish hue, slightly enlarged towards the middle, narrow at the apex, and marked off from the disk by a row of strong punctures,—and densely punctate in numerous rows, the punctures small and sub-obliterate, the interstices also sub-obliterate. but a few showing a more costiform appearance than the others. Body beneath subnitid, substriolate. Antennæ, palpi, and tarsi, of a reddish brown.

579.—PTEROHELÆUS CONFUSUS. n. sp.

Length 7 lines ; width 4 lines.

Ovate, black, subnitid. Head a little widened and elevated in front of the eyes, and scarcely emarginate in front, with the central canaliculation minute, the semi-circular suture well marked, and a transverse raised line near the apex of the clypeus. Thorax subconvex, with a broad reddish reflexed margin, and the median line scarcely traceable. Scutellum transversely and curvilinearly triangular. Elytra subconvex, with the lateral margins reddish, nearly as broad as those of the thorax at the humeral angles, and becoming narrower to the apex, with the disk covered with very numerous rows of small punctures, becoming obliterated towards the apex, and the interstices faintly costate, and quite obliterated behind. Under side of body nitid, striolate. Legs piceous. Antennæ and tarsi reddish.

580.—SARAGUS OVALIS. n. sp.

Length 9 lines.

Oblong-ovate, black. opaque. Head widened, and obtusely angled before the eyes, and broadly rounded and almost truncate in front. Thorax very deeply emarginate in front, deeply bi-emarginate at the base, and very slightly emarginate at the centre of the basal lobe, with the lateral margins very broad, reflexed and of a dull red color. Scutellum transversely triangular. Elytra of the width of the thorax at the base and of a dull chocolate color, with the lateral margins reflexed as in the thorax, broad at the humeral angles, and considerably narrowed towards the apex, and with the disk covered with rows of very small obliterate punctures. Under surface of body subnitid. Tarsi ciliated with golden yellow hair.

581.—Ospidus chrysomeloides, Pasc. *Journ. of Ent., page 468.*

582.—Nyctozoilus Mastersii. n. sp.

Length 9 lines.

Convex, oval, black, opaque, and squalid. Head coarsely punctate, subconvex, and deeply impressed on the median line, with the clypeus on a lower plane than the posterior portion of the head. Thorax transverse, not very deeply emarginate in front, and nearly truncate behind, with the lateral margins moderately broad, thick, and reflexed, the posterior angles slightly constricted on the sides and pointed backwards, and the middle of the disk vermiculate. Elytra much broader than the thorax, not longer than the width, broadly rounded at the humeral angles and the apex, and broadest in the middle, with the suture and three irregular lines on each elytron slightly but distinctly elevated, and the intervals irregularly and largely reticulate, and subfoveate. Under surface less opaque and squalid, and substriolate,

583.—Nyctozoilus elongatulus. n. sp.

Length 8 lines.

Of a more elongate and less convex form than the last species. Head and thorax densely punctate, the former depressed in the middle and in front, the latter rather longer than the width, with thick reflexed lateral margins and prominent angles. The sculpture of the elytra is the same as in *N. Mastersii*, but much more distinctly and regularly reticulate. The head is rather sharply angled in front of the eyes, and the clypeus is very slightly emarginate.

584.—Hypaulax Gayndahensis. n. sp.

Length 10 lines.

This species only differs from *H. ovalis* of Bates in having the striæ on the elytra very small, and the punctures very large.

585.—HYPAULAX OPACICOLLIS. n. sp.

Length 9 lines.

The thorax in this insect is more opaque than in the last, and the punctures on the elytra are still larger and though placed in rows, are not in striæ.

Both insects may have been described by Mr. Bates under the names of *sinuaticollis* and *tarda*, as I have never seen the descriptions of these two species.

586.—PROMETHIS PASCOEII. n. sp.

Length 9 lines.

Of the size, form, and general appearance of *Promethis angulata* Erichs, but differs in having the elytra less deeply striate, while the punctures in the striæ are much longer and less crowded.

587.—MENEPHILUS NIGERRIMUS, Boisd. *Voy. Astrol.* II., page 254.— Blanch *Voy. Pole. Sud.* IV., page 163, *t.* 11, *f.* 10.—Dej. *Cat.* 3rd. *Ed.*, page 226.—Blessig. *Hor. Ent. Ross. Soc.* 1, 1861, page 95.

Australis MacLeay Dej. *Cat.*, page 226.

588.—MENEPHILUS PARVULUS. n. sp.

Length 3 lines.

Oblong, flat, black, subnitid and finely punctate. Head broad, rounded in front of the eyes and truncate at the apex, with the terminal joint of the maxillary palpi scarcely securiform and obliquely truncate. Thorax quadrate with the anterior angles advanced, the sides parallel, and the apex truncate. Elytra of the width of the thorax, parallel-sided, rounded at the apex, and striato-punctate—the striæ small and the interstices very slightly convex.

A new genus might very properly be formed for this curious little insect.

589.—MENERISTES SERVULUS, Pasc. *Ann. Nat. Hist.*, 1869, page 151.

590.—DECHIUS APHODIOIDES, Pasc. *Journ. of Ent.*, II., *page* 445.

MICROPHYES. n. gen.

Antennæ shorter than the thorax, the third joint not quite so long as the fourth and fifth together, the 6th 7th 8th 9th and 10th joints transverse, serrate, and gradually increasing in size, the 11th nearly round. Last article of maxillary palpi obtuse and sub-cylindrical. Labrum short, transverse. Clypeus lightly emarginate, joined to the head by a semicircular suture. Eyes entirely divided by the cheeks. Thorax very transverse, emarginate in front, biemarginate behind. Body oval, subdepressed. Legs as in *Tenebrio*.

The broad oval subdepressed form of the species on which I found this genus is unlike any of the *Tenebricuidæ* proper I have hitherto come across. In other respects it approaches very nearly to *Tenebrio*.

591.—MICROPHYES RUFIPES. n. sp.

Length $2\frac{1}{2}$ lines.

Ovate, dark piceous brown, subnitid, finely and not densely punctate. Thorax twice as broad as the length, finely bordered at the sides and base, rounded on the sides, widest at the base, and broadly lobed in the middle of the apex, with the anterior angles moderately advanced. Scutellum curvilinearly triangular. Elytra as broad at the base as the base of the thorax, very slightly widened towards the apex which is round, and covered with punctures and subobsolete striæ, with faintly elevated interstices. Under side of body and legs reddish.

592.—CEPHALEUS CHALYBEIPENNIS. n. sp.

Length 8 lines.

. Broadly ovate, very convex, and very nitid. Head and thorax finely punctate, and of a golden green colour, the latter margined with purple on the sides, and impressed in the middle on the median line and on each side of it, with the anterior angles acute and reflexed. Scutellum golden green, finely punctate. Elytra much broader than the thorax, truncate at the humeral angles, very convex, of a beautiful chalybeate blue colour, with golden green margin, and marked in addition to the general fine punctuation, with several irregular rows of very large rather distant punctures. Antennæ, palpi, legs, and under surface of body black and nitid.

593.—CYPHALEUS CUPRICOLLIS. n. sp.

Length 7 lines.

Head golden green, densely punctate. Thorax of a brilliant coppery red, lightly punctate, and almost angled in the middle of the sides, with the anterior angles obtuse. Elytra convex, broader than the thorax, rounded at the humeral angles, rather broader behind than in front, very densely and obliterately punctate, and of a subnitid bluish green colour, with a large coppery patch on the sides at the apex. Legs and under side of body greenish black, and nitid.

594.—PROPHANES WESTWOODII. n. sp.

Length 12 lines.

Elongate-oval, slightly convex, finely punctate, and of a nitid bronzy black, with a short brown pubescence. Head flat, transversely impressed, with the clypeus emarginate. Thorax with the anterior angles long, acute, and reflexed at the tip. Elytra of an olive colour, confusedly punctate, and terminating with a short slightly recurved spine. Legs and under side black. Antennæ piceous.

This species has a general resemblance to *P. aculeatus*, Westw.

595.—CHÆTOPTERYX MASTERSII. n. sp.

Length 6 lines.

Ovate, convex, punctate, nitid, and clothed with long erect hairs. Head and thorax brassy, the latter having a ruddy gloss, with the anterior angles short and not acute, and a broad depression near the posterior angles. Elytra broader than the

s

thorax, a little enlarged behind, terminating in a minute spine, covered with large punctures, and of the most varied splendour of colouring, being golden on the suture, of a ruby colour next, then metallic green, and reddish purple towards the sides behind. Antennæ, legs and body beneath, black.

596.—ATRYPHODES OPACICOLLIS.   n. sp.

Length 9 lines.

Elongate, black, opaque. Head marked with a stirrup shaped impression in front. Thorax longer than the width, narrower at the base than in front, and emarginate at both apex and base, with the anterior angles rather prominent and subacute, the sides gradually rounded to near the base, then subabruptly narrowed until close to the posterior angles when they become straight and make with the emargination of the base the posterior angles acute, and with a distinct median line and a broad sublateral depression giving the appearance of a broad margin. Elytra rather narrower than the thorax, subangular and somewhat reflexed at the humeral angles, marked with ten deep striæ counting the lateral one on each elytron, and with the interstices convex and of equal size. Under surface subnitid. Tarsi piceous.

597.—ATRYPHODES PITHECIUS, Pasc.   Ann. Nat. Hist., 1869, page 39.

598.--ATRYPHODES MASTERSII.   n. sp.

Length 6½ lines.

Of a bronzy olive, subnitid. Head marked with a stirrup shaped impression. Thorax a little broader than the length, emarginate in front, very slightly so at the base, a little narrower at the base than at the apex, with the sides gradually rounded, the anterior angles advanced and subobtuse, the posterior angles acute, the median line well marked, the sublateral depressions moderately so leaving a rather broad marginal space in the middle, and the base deeply impressed near the posterior angles. Elytra rounded at the humeral angles and marked with eight deep striæ on each elytron,—the lateral striæ lightly punctate,—and with the interstices subconvex. Under surface black, nitid. Tarsi piceous.

599.—ADELIUM STRIATUM, Pasc. *Journ. of Ent.* II., *page* 481.

600.—ADELIUM VIRIDIPENNE. n. sp.

Length 10 lines.

This species very closely resembles *A. striatum*, Pasc., and is perhaps only a variety of that insect. It is more nitid, without foveæ in the middle of the thorax, and has the clytra entirely of a blackish-green colour.

601.—ADELIUM RUGOSICOLLE. n. sp.

Length 8 lines.

Resembles *A. plicigerum*, Pasc. The thorax however is more punctate, and vermiculately rugose, and the whole is of a more decided copper colour.

602.—ADELIUM AUGURALE, Pasc. *Journ. of Ent.* II., *page* 480.

603.—ADELIUM SCUTELLARE, Pasc. *Ann. Nat. Hist.*, 1869, *page* 134.

604.—ADELIUM REPANDUM, Pasc. *Ann. Nat. Hist.*, 1869, *page* 137.

605.—ADELIUM CONVEXIUSCULUM. n. sp.

Length 6 lines.

Of a bronzy olive, nitid, convex, and ovate. Head finely punctate, and a little rounded in front, with the suture of the epistome semicircular. Thorax transverse, not broader at the base than at the apex, finely punctate, foveate near the sides, and subrugose, with the sides rounded, and the posterior angles acute and slightly directed outwards. Elytra broader than the thorax and not much longer than the width, with nine very fine punctate striæ on each clytron, the punctures mostly small but sometimes rather large and elongate, the interstices smooth. Under side of body black. Tarsi piceous.

This insect is. I imagine, somewhat like *A. ancilla*, Pasc., a species I have never seen. The sculpture however appears to be different, and in the size of the antennæ there must be a very marked difference.

606.—ADELIUM GEMINATUM, Pasc. *Ann. Nat. Hist., February*, 1870.

607.—ADELIUM REDUCTUM, Pasc. *Ann. Nat. Hist.,* 1869, *page* 135.

608.—ADELIUM PARVULUM. n. sp.

Length 4 lines.

Elongate-ovate, subconvex, black, subnitid, and densely punctate. Head slightly rounded in front, with the epistome a little advanced over the labrum, and with the third joint of the antennæ not much longer than the fourth. Thorax scarcely broader than the length, and not broader behind than in front, with the apex lightly emarginate, the sides moderately rounded, the base truncate, and the median line distinctly marked. Scutellum transversely triangular. Elytra scarcely broader than the thorax at its broadest part, of a slightly bronzy hue and striato-punctate, with the interstices subconvex and finely punctate. Body beneath minutely punctate. Tarsi piceous.

Along with this insect I find in the collection one, which though differing considerably, may possibly be the male of the same species. It is smaller, is without the well marked median line on the thorax of the other, and has the striæ of the elytra more profound and the interstices more convex.

609.—ADELIUM PANAGÆICOLLE. n. sp.

Length 4 lines.

Elongate-ovate, subdepressed, black, subopaque, densely and coarsely punctate. Head short, and truncate in front, with the suture lightly marked. Thorax much like that of the genus

Panagæus, transverse, marked on the median line, and slightly emarginate in front, with the sides much widened into a rounded angle in the middle, and emarginately narrowed from there to the posterior angles which are acute and very slightly pointed outwards. Scutellum very transversely triangular, smooth at the apex. Elytra little broader than the thorax and striate, the striæ becoming punctate and sometimes interrupted towards the sides, with the interstices subconvex and finely punctate. Body beneath subnitid and very finely punctate. Antennæ and tarsi piceous, the former widening and subserrate towards the apex.

610.—Adelium monilicorne. n. sp.

Length 3 lines.

Elongate-ovate, subconvex, black, subnitid, and punctate. Head coarsely punctate, slightly rounded in front, and rather prominently and obtusely angled in front of the eyes, with the suture of the epistome subsemicircular. Thorax transverse, not wider behind than in front, and rounded on the sides. Elytra scarcely broader than the thorax at its broadest part and profoundly striato-punctate, with the interstices convex, and minutely punctate. Body beneath finely punctate. Antennæ and tarsi piceous red, the former short, submoniliform, and becoming larger towards the apex, with the terminal joint round.

611.—Seirotrana Mastersii, Pasc. *Ann. Nat. Hist.*, *February*, 1870.

612.—Seirotrana punctifera. n. sp.

Length 8 lines.

Of a bronzy olive, subnitid. Head deeply impressed on the suture, subtruncate. Thorax subtransverse, slightly rounded on the sides, trifoveate at the base, and marked on the disk with a few large scattered punctures. Scutellum very transverse, broadly rounded behind. Elytra finely striato-punctate with the interstices flat, the 3rd 5th and 7th marked with a few large punctures. Body beneath black. Antennæ and tarsi piceous.

613.—SEIROTRANA NOSODERMOIDES, Pasc. *Ann. Nat. Hist., February*, 1870.

614.—SEIROTRANA FEMORALIS. n. sp.

Length 3¾ lines.

Coppery brown, subopaque, punctate and flat. Head and thorax densely punctate, the latter subquadrate, slightly emarginate at the apex, and rounded on the sides. Scutellum transverse, rounded behind, of a greenish colour, nitid and punctate. Elytra with four fine very interrupted costæ on each, with two rows of large rather distant punctures between these costæ, and with a few distant granules between the rows. Legs brown with the apex of the thighs yellow and the tarsi reddish.

615.—CORIPERA MASTERSII. n. sp.

Length 5 lines.

Dark copper-brown, nitid, flat. Head coarsely punctate, subemarginate in front. Thorax subquadrate, emarginate in front, lightly rounded on the side, and marked on the disk with a few large punctures, with the median line distinct and the posterior angles showing a flat square surface. Scutellum transversely triangular. Elytra marked with a sutural and three double striæ strongly punctate, and with three broad flat interstices marked with a number of elongate punctate depressions placed in pairs, these being more numerous in the first and third interstice than in the second.

The character of the sculpture in this species, is, though much more marked, somewhat like that of *C. ocellata*, Pascoe.

616.—LICINOMA ELATA, Pasc. *Ann. Nat. Hist., February*, 1870.

617.—LICINOMA VIOLACEA. n. sp.

Length 5½ lines.

Elongate, black, subnitid. Head roughly punctate, subrugose. Thorax subquadrate, very finely punctate, truncate in front and behind, widest in the middle, and a little narrower at the base than at the apex. Elytra of a violet huc and punctato-striate, with the interstices broad and subconvex. Body beneath piceous-brown. Tibiæ and tarsi reddish.

618.—BRYCOPIA LONGIPES. n. sp.

Length 3 lines.

Bronzy black, nitid, punctate. Head with the suture curved, and without lateral impressions. Thorax subquadrate, truncate in front and behind, rounded on the sides, and considerably narrower at the base than at the apex, with two faintly marked foveæ near the middle. Scutellum triangular. Elytra deeply striato-punctate, with the interstices subconvex and finely punctate. Under surface and thighs brown. Antennæ, tibiæ and tarsi reddish. Hind legs long.

619.—BRYCOPIA DUBIA. n. sp.

Length 2½ lines.

Ovate, subconvex, black, and nitid. Head with the suture straight, and with the lateral canals at right angles to it. Thorax transverse, slightly emarginate in front, rounded a little on the sides, and very thinly and minutely punctate, with the posterior angles acute and minutely recurved. Elytra obovate, broader than the thorax, and striato-punctate with the interstices smooth and nearly flat. Body beneath brown, nitid. Antennæ and legs piceous.

I am perhaps wrong in placing this insect in the genus *Brycopia*. It seems in some respects to show most affinity to *Adelium monilicorne* described by me some pages back. I think that a new subgenus might well be formed for the reception of both.

LEPTOGASTRUS. n. gen.

Antennæ, thick, and of the length of the head and thorax, with the third joint little longer than the fourth, the other joints increasing gradually in width up to the eleventh, which is very large and oval. Thorax elongate, narrowed behind. Elytra elongate-oval. Thighs robust. Body pedunculate, general form narrow, subcylindrical. In other respects resembling *Adelium*.

The elongate form, pedunculated body, and clavate antennæ, are the most marked characteristics of this genus, and separate it widely from all others of the group. The genus *Licinoma* Pascoe is the one perhaps to which it approaches nearest.

620.—LEPTOGASTRUS MASTERSII. n. sp.

Length 2⅔ lines.

Dark copper-brown, subnitid, punctate. Head coarsely punctate, with the epistome subemarginate, the suture deeply marked, and without lateral canals. Thorax truncate at the apex. not rounded at the sides, and much narrowed at the base which is truncate. Elytra not broader than the thorax, of an elongate-oval shape and deeply striato-punctate. Thighs yellow on the apical half. Tibiæ and tarsi reddish.

621.—OMOLIPUS CORVUS, Pasc. *Journ. of Ent.*, I., *page* 127, *t.* 6, *f.* 9 ; *Ann. Nat. Hist.*, 1869, *page* 143.

622.—OMOLIPUS GNESIOIDES, Pasc. *Ann. Nat. Hist.*, 1869, *page* 143.

623.—OMOLIPUS GRANDIS. n. sp.

Length 8 lines.

Convex, black, subopaque, with the suture of the epistome more deeply impressed than in *O. corvus*. Thorax more opaque. Elytra punctate in the same way as *corvus*, but the punctures larger and less acutely impressed.

624.—AMARYGMUS RUFIPES. n. sp.

Length 6 lines.

Oblong-ovate. Head, thorax and scutellum black, opaque, minutely punctate. Elytra of a semiopaque silky blue, with eight regular rows of small well marked punctures on each elytron. Antennæ piceous. Legs entirely red. Body beneath black, subnitid.

The species named by Mr. Pascoe *nigritarsis*, is the most like to this of all the described species.

625.—AMARYGMUS PICIPES.   n. sp.

Length 5½ lines.

Shorter and broader than the last, with the elytra green, sub-nitid, and a little more largely punctured than in the former species, and the legs entirely of a piceous brown. In other respects like *A. rufipes*.

626.—AMARYGMUS OPACICOLLIS.   n. sp.

Length 6 lines.

Oblong-ovate. Head black, punctate, and flat between the antennæ, with the eyes prominent and not covered by the thorax. Thorax brassy black, opaque, and minutely punctate. Elytra much broader than the base of the thorax, of a purplish blue colour be-coming green towards the sides, subnitid, and marked on each elytron with eight rows of small closely placed subelongate punctures. Under side of body, antennæ and legs, dark brown and subnitid.

627.—AMARYGMUS GRANDIS.   n. sp.

Length 9 lines.

Oblong-ovate. Head black, punctate, with the eyes large and approximate. Thorax short, narrow in front, coppery black, sub-opaque, and very minutely punctate. Scutellum black, triangular, and smooth. Elytra cyaneous at the suture and showing green, purple, blue, and coppery red reflexions over the rest of their sur-face, with regular rows of small punctures, and the interstices minutely and somewhat rugosely punctate. Under surface black, subnitid, minutely punctate and striolate. Legs and antennæ brownish-black.

628.—AMARYGMUS CUPREUS, Fabr. *Syst. Ent.*, *page* 123 ; *Syst. El.* II., *page* 12 ;—Oliv. *Ent.* III., 58, *page* 7, *t.* 1, *f.* 6.

I am not quite positive as to the identity of this insect.

629.—AMARYGMUS RUGOSICOLLIS.   n. sp.

Length 8 lines.

Oblong-ovate. Head black, and densely punctate, with the

eyes large and partially covered by the thorax. Thorax short, not much narrower in front than behind, of a reddish purple, opaque, densely punctate, and somewhat rugose in the middle. Scutellum purplish black, thinly punctate, and curvilinearly triangular. Elytra nitid, viridi-æneous on the suture, and coppery red and metallic green on the rest of their surface, with eight rows on each elytron of large deeply impressed punctures and with the intersticcs smooth and minutely punctate. Body beneath and legs, black, subnitid, and finely punctate.

630.—AMARYGMUS PUNCTIPENNIS. n. sp.

Length 6 lines.

Oblong-ovate. Head black, and rather thinly punctate, with the eyes partially covered by the thorax. Thorax of a coppery green, subopaque, and finely but not densely punctate. Scutellum greenish black, triangular, the sides a little rounded. Elytra very nitid, viridi-æneous with the suture and a broad median vitta of a purplish red, and marked with eight regular rows of strong punctures—larger and more distant than in the last species—on each elytron, and with the interstices smooth and minutely and thinly punctate. Under surface of body and legs black, subnitid, the former very finely striolate.

631.—AMARYGMUS OBSOLETUS. n. sp.

Length 6 lines.

Ovate. Head black, and punctate, with the eyes almost completely covered by the thorax. Thorax short, broad, not much narrowed in front, green clouded with coppery red, subopaque, and finely punctate. Scutellum black, smooth, curvilinearly triangular. Elytra more convex than in most species, green with numerous patches of coppery red, subopaque, and punctate, with the usual rows of punctures almost if not quite obsolete. Under surface and legs black, nitid, finely and thinly punctate.

632.—AMARYGMUS RUGOSIPENNIS. n. sp.

Length 8 lines.

Elongate-ovate. Head black, and punctate, with the eyes

partially covered. Thorax greenish black with a dull reddish reflexion, subopaque, and finely and thinly punctate. Scutellum black, triangular, and impressed towards the apex. Elytra not much wider than the base of the thorax, nearly three times as long as the width, nitid, of a green colour with purplish red reflexions, and marked with eight rows of punctures on each elytron, and with the interstices so coarsely punctate as to make these rows appear indistinct. Under side of body and legs black, nitid, and finely punctate, with the sides of the abdominal segments deeply impressed. The antennæ and legs of this species are unusually robust.

633.—AMARYGMUS CONVEXUS, Pasc.   *Journ. of Ent.*
II., *page* 485.

634.—AMARYGMUS FOVEOLATUS.   n. sp.

Length 4 lines.

Ovate, convex. Head black, thinly punctate, and deeply impressed on the suture, with the eyes almost completely covered by the thorax and not approximate. Thorax black, subnitid, very minutely and thinly punctate. Scutellum transverse, black, and curvilinearly triangular. Elytra dark green, nitid, and marked with eight rows of foveolate punctures on each elytron. Body beneath black, subnitid. Tarsi piceous.

635.—AMARYGMUS STRIATUS.   n. sp.

Length 3½ lines.

Ovate, convex, black, subopaque. Head thinly punctate, eyes rather distant. Thorax minutely and thinly punctate, and slightly impressed transversely near the base. Scutellum curvilinearly triangular. Elytra punctato-striate, with the interstices broad and subconvex. Under side of body, antennæ and legs dark piceous.

636.—AMARYGMUS CONVEXIUSCULUS.   n. sp.

Length 3 lines.

Elongate-ovate, convex, subopaque. Head black, and deeply

impressed on the suture of the epistome, with the eyes large, and subapproximate in front. Thorax black, very minutely punctate. Elytra of a bluish black, and punctate in rows, the punctures distinct, but not so large as in *A. convexus*. Body beneath piceous black. Legs and antennæ red.

637.—Strongylium Mastersii. n. sp.

Length 5½ lines.

Black, subnitid. Head punctate, with the epistome more finely and densely punctated and with the suture very profoundly impressed. Thorax nearly square, a little rounded at the anterior angles, and finely punctate, with a strong marginal fold at the base. Scutellum finely punctate on the sides. Elytra elongate, cylindrical, broader and more nitid than the thorax, and slightly striate, with a regular row of distinct subapproximate punctures in each stria. Body beneath and legs piceous, black, nitid. Tarsi and terminal joint of antennæ ferruginous.

638.—Strongylium ruficolle. n. sp.

Length 4 lines.

Elongate, dark red, subnitid and punctate. Thorax subtransverse, scarcely rounded on the sides, truncate in front and behind, and rounded at the anterior angles, with the median line lightly marked, and with a raised basal margin. Elytra a little wider than the thorax, black, and profoundly striato-punctate.

CISTELIDÆ.

639.—Apellatus palpalis. n. sp.

Length 2¾ lines.

Pale red, subnitid, punctate and covered with a fine pubescence. Thorax subquadrate, broader behind than in front and truncate at the base, with the median line deeply impressed behind and with a small fovea on each side between it and the posterior angles. Scutellum large and rounded behind. Elytra broader than the thorax and of a paler colour, punctate and striato-punctate, with the apex, the sides and the suture, black. Legs pale. Antennæ with the middle joints rather broad and flattened. Terminal joint of maxillary palpi transversely elongate.

640.—APELLATUS MASTERSII.   n. sp.

Length 3½ lines.

This species is larger than the last, and of a darker red colour, with the maxillary palpi less elongate, the thorax proportionally less long and more rounded at the anterior angles, and with the elytra more deeply striato-punctate, and entirely black excepting a testaceous patch at the base.

641.—METISTETE PASCOEI.   n. sp.

Length 6 lines.

Black, nitid. Head and thorax finely punctate, the latter small, subquadrate, rounded at the anterior angles, and scarcely broader at the base than at the apex. Elytra much broader than the thorax and broadest near the apex, subacuminate at the apex, three times longer than the width, and sharply striato-punctate. Legs pale red. Antennæ, palpi, and labrum, piceous red.

642.—ATRACTUS RUFICOLLIS.   n. sp.

Length 4 lines.

Red, nitid, and thinly punctate. Thorax longer than the width, a little narrowed in front, truncate at the apex and base, and rather deeply impressed at the base of the median line. Elytra of a brilliant purplish blue, and striato-punctate. Legs and prothorax red. Tarsi brown. Abdomen, meso- and meta-thorax black. Antennæ dark brown, joints 4 to 10 broad and subserrate.

This species and the following ought to constitute a new genus. I place them at present with *Atractus*, because it is not improbable that the genus *Licymnius* of Bates, of which I have never seen the description, may have been made for one of these very species.

643.—ATRACTUS CYANEUS.   n. sp.

Length 3½ lines.

This species only differs from the last in being smaller, in having the head and thorax of a bronzy black, and in having the legs brown and finely pubescent.

644.—ATRACTUS VITTICOLLIS. n. sp.

Length 4½ lines.

Red, subnitid, punctate and clothed with a very short pale pubescence. Head broad, and not narrowed behind the eyes, with a black spot on the forehead. Thorax not broader than the head, longer than the width, and not narrowed in front, with a central black vitta. Scutellum black, and rounded behind. Elytra broader than the thorax, and rugosely striato-punctate, with a purplish gloss in the middle. Abdomen and legs piceous brown.

645.—ATRACTUS RUGOSULUS. n. sp.

Length 4 lines.

Piceous black, subnitid, coarsely punctate and pubescent. Thorax of the same form as in the preceding species, but of a bronzy black, with the median line distinct. Elytra dark red and very rugosely striato-punctate, with a slight greenish tinge on the suture.

This and the species before it, *A. viticollis*, might also I think be separated from *Atractus*.

646.—CHROMOMÆA MASTERSII. n. sp.

Length 3¼ lines.

Bronzy black, nitid, and punctate. Thorax longer than the width, and not narrowed in front, with the median line deeply impressed at the base. Scutellum black. Elytra red and profoundly striato-punctate. Legs red. Antennæ brown.

647.—CHROMOMÆA PICEA. n. sp.

Length 5 lines.

Dark piceous with the elytra and legs of a redder hue, nitid, and finely punctate. Thorax scarcely longer than the width, and slightly rounded at the anterior angles, with a shallow fovea near the middle of each side, and the median line slightly impressed at the base. Elytra much broader than the thorax and finely striato-punctate,—the interstices broad and flat. Abdomen black, finely pubescent.

648.—HOMOTRYSIS RUFICORNIS. n. sp.

Length 8 lines.

Black, subnitid, and punctate. Head and thorax finely punctate, the latter subtransverse, rounded and narrowed at the anterior angles and truncate at the base Elytra broad, convex, broadest behind the middle, more than twice longer than the width, and covered with large punctures, with 8 distinct punctured striæ on each elytron, and a short one near the scutellum running into the first. Antennæ red. Legs nitid, with the middle of the tibiæ red.

649.—HOMOTRYSIS SUBGEMINATUS. n. sp.

Length 7¼ lines.

Black, subnitid and punctate. Head and thorax finely and thinly punctate, the latter transverse and much broader at the base than at the apex. Elytra broad, convex, broadest behind the middle, densely punctate and very lightly marked with subgeminate punctate striæ. Legs and antennæ black.

650.—HOMOTRYSIS REGULARIS. n. sp.

Length ♂ 4 lines, ♀ 7 lines.

Black, subnitid. Head and thorax densely and finely punctate, the latter subtransverse and very little broader at the base than at the apex, with the median line lightly marked on its anterior half. Elytra very finely punctate, and striato-punctate, the punctures in the striæ being large at the base and small towards the apex.

651.—ALLECULA ELONGATA. n. sp.

Length 7½ lines.

Black, subnitid, finely punctate and clothed with a short, erect, brown pubescence. Thorax broad, subconvex, subquadrate, slightly lobed at the apex, and slightly bi-emarginate and bi-impressed at the base, with the anterior angles rounded and the posterior acute. Elytra a little broader than the thorax, nearly three times longer than the width, densely and finely punctate, and marked with very fine striæ, and a series of large square punctures in a groove extending from the humeral angles to beyond the middle.

652.— ALLECULA SUBSULCATA. n. sp.

Length 5½ lines.

Dark brown, and subnitid, with the parts of the mouth, apex of the antennæ, and tarsi, piceous red. Head and thorax densely punctate, the latter not broader than the head, almost square, a little rounded on the sides and broadly depressed on the anterior half of the median line. Elytra deeply striato-punctate, with the interstices smooth and slightly elevated, and with an abbreviated stria near the scutellum.

653.—ALLECULA PUNCTIPENNIS. n. sp.

Length 6½ lines.

Brown, opaque, and clothed with short semi-erect brown setæ. Head and thorax densely and finely punctate, the latter rather broader than the head, subquadrate, slightly rounded on the sides. and without trace of median line. Elytra a little broader than the thorax and deeply striato-punctate, with the punctures larger towards the base, and with the interstices slightly elevated. Legs and palpi opaque, reddish brown.

654.—ALLECULA PASCOEI. n. sp.

Length 5½ lines.

Brown, opaque, densely and finely punctate, and clothed with a fulvous pubescence. Thorax transverse, rounded on the sides and anterior angles, and slightly lobed at the base. Elytra broader than the thorax, and in addition to the dense puncturation of the whole surface, marked with fine but distinct punctate striæ with an abbreviate one near the scutellum. Under surface, legs, and antennæ of a reddish brown.

655.—ALLECULA MASTERSII. n. sp.

Length 3½ lines.

Brown, subopaque, densely punctate and densely clothed with

a short semi-erect yellowish pubescence. Thorax transverse, rounded at the anterior angles, and broader at the base than at the apex. Elytra closely and coarsely punctate and marked with fine striæ rather indistinct towards the base. Antennæ, palpi, and tarsi, reddish. Under surface of body piceous and nitid.

656.— ALLECULA PLANICOLLIS. n. sp.

Length 3½ lines.

Brown, subnitid, punctate and very finely pubescent. Thorax quadrate, slightly rounded on the sides anteriorly, broadly depressed and flattened on the disc, and bi-foveate at the base. Elytra a little broader than the thorax, densely and rather confusedly punctate, and rather profoundly striato-punctate. Body beneath, legs, antennæ, and the parts of the mouth, ferruginous.

657.— CISTELA CONVEXA. n. sp.

Length 3¼ lines.

Oval, convex, of an olive-brown colour, subnitid, densely and very minutely punctate, and densely clothed with a short whitish pubescence. Thorax very transverse, rounded at the anterior angles, much broader behind than in front, and slightly bi-foveate at the base. Elytra lightly striate with the interstices sub-elevated. Body beneath and legs piceous-black and nitid.

658.— CISTELA OVATA. n. sp.

Length 2½ lines.

This species differs from the last in being much smaller and less convex, in being of a pale reddish brown, and in having the striæ of the elytra almost obsolete.

659.— CISTELA DEPRESSIUSCALA. n. sp.

Length 2½ lines.

Ovate, subdepressed, piceous brown, subnitid, densely punctate, and densely clothed with a fulvous pubescence. Thorax very transverse. Elytra rather coarsely punctate, and striato-punctate. Legs red.

T

660.—CISTELA POLITA. n. sp.

Length 2 lines.

Elongate-ovate, red, nitid and punctate. Thorax transverse, rounded on the sides, and slightly broader at the base than at the apex, with the median line and two foveæ traceable at the base. Elytra black, very nitid, and striato-punctate, with the interstices smooth and flat. Legs pale red.

MELANDRYIDÆ.

661.—ORCHESIA ELONGATA. n. sp.

Length 3 lines.

Elongate, reddish brown, subopaque, densely and minutely punctate, and covered with a fulvous sericeous pubescence. Legs, antennæ, and palpi of a paler colour.

LAGRIIDÆ.

662.—LAGRIA GRANDIS, Gyllh.   *Schonh. His.* 1, *Ins. app.*, 3, *page* 9.—Blanch. *Voy. Pole. Sud.* IV., *page* 183, *t.* 12, *f.* 9.—Erichs. *Weigm. Archiv.* 1842, I., *page* 370.
     rufescens Boisd. *Voy. Astrol.* II., *page* 286, —Latr. *Dej. Cat.*, *3rd ed.*, *page* 237.
     ruficollis MacLeay *Dej. Cat.* 3rd ed., *page* 237.

663.—LAGRIA CYANEA. n. sp.

Length 3¾ lines.

Of a greenish blue with a tinge of purple on the elytra, moderately convex, subopaque, densely punctate and cinereo-pilose. Under surface of body, coxæ, basal portion of the femora and basal joints of antennæ red, rest of antennæ, and legs black.

OMMATOPHORUS. n. gen.

Head small. Neck distinct. Eyes large, round and contiguous. Maxillary palpi securiform. Antennæ long, filiform,

1st joint thick and clavate, 2nd short, 3rd longer, the rest gradually increasing in length to the apical one which is very long. Thorax flat, transverse and truncate in front and behind. Elytra broader than the thorax, flat, parallel-sided and rounded at the apex. Legs as in *Lagria*.

664.—OMMATOPHORUS MASTERSII. n. sp.

Length 2½ lines.

Dark red, subnitid, coarsely punctate and clothed with black hair. Head black. Thorax rounded at the anterior angles, and square behind. Elytra deeply striato-punctate, and of a brownish colour excepting on the sides and suture. Under side of body piceous. Legs pale red.

This is a very distinct genus, but I may be wrong in classing it with the *Lagriidæ*.

ANTHICIDÆ.

665.—MECYNOTARSUS KREUSLERI, King.   *Trans. Ent. Soc. N. S. Wales*, II., *page* 4.

666.—MECYNOTARSUS KINGII. n. sp.

Length 1¼ lines.

Much resembling *M. concolor* King. The thorax is very convex, of a brownish colour and densely covered with a silvery pubescence, the elytra are more red and clothed with a cinereous subsericeous pubescence. The legs and antennæ are red.

667.—MECYNOTARSUS MASTERSII. n. sp.

Length 1½ lines.

This species very much resembles *M. Kreusleri* King, it is however a larger and more beautiful insect. The form is the same, but the thorax is covered with a dense silky olive pubescence, and there is a large triangular patch of the same on the elytra in the scutellar region, which interrupts the white sub-basal fascia in the middle. The apex is also of a sericeous olive hue. In all else it is the same as *M. Kreusleri*.

668.—FORMICOMUS KINGII. n. sp.

Length 2 lines.

Black, sub-opaque, very densely punctate and pubescent. Thorax longer than the width, broad and rounded near the front, and narrowed at the base. Elytra elongate-oval, subconvex, subnitid, with two narrow fasciæ not reaching the suture, composed of silvery pubescence. Hind thighs very large and strongly toothed on the under side near the apex. Antennæ, base of thorax, scutellum, base of femora, tibiæ and tarsi, piceous red.

669.—FORMICOMUS DENISONII, King. *Trans. Ent. Soc. N. S. Wales*, II., *page* 6.

670.—FORMICOMUS HUMERALIS. n. sp.

Length 1½ lines.

Head and thorax red, and subnitid, the latter narrower than the head, convex, much narrowed at the base and deeply impressed on the median line. Elytra broadly ovate, convex, black, nitid, and thinly clothed with hairs, with an oblique white fascia commencing near the humeral angle, and not reaching the suture. Legs and terminal joints of antennæ black.

671.—ANTHICUS LURIDUS, King. *Trans. Ent. Soc. N. S. Wales*, II., *page* 16.

672.—ANTHICUS PULCHER, King. *Trans. Ent. Soc. N. S. Wales*, II., *page* 12.

673.—ANTHICUS KINGII. n. sp.

Length 1½ lines.

Black, subnitid, punctate, and hairy. Thorax narrower than the head, rounded in front, subconvex, and compressed laterally behind. Elytra broad, subdepressed, densely and coarsely punctate, and of a dark red colour with a broad black median fascia not reaching the suture. Legs and antennæ red.

674.—ANTHICUS PROPINQUUS. n. sp.

Length 1½ lines.

Piceous red, subnitid, punctate and hairy. Thorax narrower than the head, convex and scarcely compressed laterally behind. Elytra subovate, convex, and sparsely and coarsely punctate, with the base, apex, and a median fascia not reaching the suture, black. Antennæ and legs pale red.

675.—ANTHICUS LATICOLLIS. n. sp.

Length 1½ lines.

Head transverse, black. Thorax transverse, subconvex, as broad as the head and much narrowed behind, of a dark red colour and clothed with a cinereous pubescence. Elytra red, pubescent, subconvex, subnitid, and finely punctate, with a large median fascia becoming narrower towards the suture but not reaching it, and the apex, black. Antennæ and legs red, the latter with the apical half of the thighs brown.

676.—ANTHICUS MASTERSII. n. sp.

Length 1½ lines.

Black, nitid, and thinly punctate. Thorax subconvex, narrower than the head, longer than the width, slightly and gradually narrowed towards the base, and piceous at the base. Elytra subconvex, with two indistinct deep red transverse spots on each elytron, one near the base, the other behind the middle. Antennæ and legs piceous.

677.—ANTHICUS CONSTRICTUS. n. sp.

Length 1¼ lines.

Black, nitid, minutely punctate, and sparingly pubescent. Thorax red, elongate, narrower than the head and much constricted in the middle. Elytra flat, with a narrow yellow fascia near the base. Antennæ and legs reddish.

678.—ANTHICUS PALLIDUS. n. sp.

Length 1 line.

Pale red, subnitid, densely punctate and finely pubescent.

Thorax nearly as broad as the head, not longer than the width, and narrowed gradually to the base. Elytra somewhat depressed, and clouded with brown towards the apex and sides. Eyes very small.

679.—ANTHICUS ABERRANS. n. sp.

Length 2 lines.

Elongate, pale reddish brown, opaque, punctate and fulvo-pubescent. Thorax ovate, and as broad as the head, with the median line faintly marked towards the base. Elytra elongate, parallel-sided, subdepressed, scarcely wider than the thorax and of a more red colour, and striato-punctate. Legs and antennæ pale red, the latter having the 9th and 10th joints broader than the others, and the last joint long.

This species ought to constitute a separate genus.

PYROCHROIDÆ.

680.—LEMODES MASTERSII. n. sp.

Length 2 lines.

This species differs from *Lemodes coccinea* Bohem, in being of a smaller size, broader form, and duller colour and in having the legs and basal joints of the antennæ red, and the two apical joints white.

MORDELLIDÆ.

681.—MORDELLA OCTOMACULATA. n. sp.

Length 4 lines.

Satiny black, opaque. Head silvery white in front and behind. Thorax bordered with white and with a white fascia interrupted in the middle. Elytra with an oblique oval spot in the middle near the base, a straight oval spot behind near the suture, a round spot towards the apex and a transverse spot on the side towards the shoulders, white. Body beneath spotted with white.

682.—MORDELLA 14 MACULATA. n. sp.

Length 4 lines.

Of a more elongate form than the last species. The fascia on the thorax is narrow, wavy, and complete, and there are two small spots behind it. The spots on the elytra are small and are placed one at the scutellum, one a little way from the base near the centre, one about the middle near the suture, one between that and the apex, one very small on the side near the apex, and two also small behind the shoulder. The markings on the under side are much the same as in *M. octomaculata*.

683.—Mordella aterrima. n. sp.

Length 3 lines.

Elongate, satiny black, with a white spot on the side of the basal segment of the abdomen.

684.—Mordella Australis, Boisd. *Voy. Astrol.* II., *page* 289.

685.—Mordella brunneipennis. n. sp.

Length 2 lines.

Narrow, black, opaque. Elytra chesnut brown with the suture narrowly margined with black.

686.—Mordella cuspidata. n. sp.

Length 1¼ lines.

Head, thorax, base of elytra, under surface of thorax, and legs, red, all the rest of the elytra and the abdominal segments, black. Anal spine long and very acute.

RHIPIPHORIDÆ.

687.—Trigonodera Gerstäckeri. n. sp.

Length 6½ lines.

Brown, subopaque, densely and very minutely punctate and clothed with a fine sericeous fulvous pubescence. Scutellum oblong, rounded behind. Elytra moderately attenuated towards the apex and marked with obsolete traces of costæ. Under surface and legs reddish brown and clothed with a very fine cinereous pubescence. Antennæ with the first four joints simple, the third longest.

688.—TRIGONODERA MASTERSII. n. sp.

Length 4½ lines.

Of more elongate form than the last, dark brown, holosericeous, and subopaque. Thorax rather longer than the width. Elytra marked with numerous spots of a deeper brown, giving an appearance of indistinct fasciæ on the apical half. Under surface and legs reddish brown and clothed with a very fine whitish pubescence. Antennæ red, third joint long.

689.—PTILOPHORUS GERSTÄCKERI. n. sp.

Length 3 lines.

Elongate, brown, opaque, very densely and minutely punctate, and densely clothed with very short cinereous pubescence. Thorax convex and profoundly bi-emarginate at the base. Scutellum large, subtriangular and rounded at the apex. Elytra reddish-brown, and having a patchy appearance from the cinereous pubescence not covering equally the whole surface. Antennæ largely pectinated in the male.

690.—RHIPIPHORUS LUTEIPENNIS. n. sp.

Length 2 lines.

Black, nitid, and thinly punctate. Elytra pale luteous and acute at the apex. Abdomen and base of tibiæ red. Antennæ red, pecten brown.

CANTHARIDÆ.

691.—ZONITIS LUTEA. n. sp.

Length 5 lines.

Upper surface entirely luteous, under surface, legs, and antennæ black. Head and thorax nitid and thinly punctate, the latter scarcely longer than the width, and faintly marked on the median line. Elytra subnitid, densely punctate, and finely pubescent.

692.—ZONITIS FUSCICORNIS. n. sp.

Length 4 lines.

Pale luteous, subnitid. Head and thorax thinly punctate, the latter elongate, not much broader behind than in front, and distinctly impressed on the basal part of the median line. Elytra subrugosely and densely punctate and pubescent. Antennæ, apex of thighs, tibiæ and tarsi, brown. Abdominal segments dusky.

693.—Zonitis apicalis. n. sp.

Length 5 lines.

Luteous, subnitid, very sparingly punctate. Thorax not longer than the width, much narrowed at the apex, rounded on the sides, and narrowed towards the base. Elytra subdepressed, broad, and broadest at the apex, which is black. Antennæ, apex of thighs, tibiæ and tarsi, black. Abdomen dusky brown.

694.—Zonitis bizonata. n. sp.

Length 5 lines.

This species differs only from the last in the elytra which are in this insect more convex, very finely and densely punctate, and broadly fasciated with black at the base and apex.

695.—Zonitis annulata. n. sp.

Length 4½ lines.

Pale luteous, subnitid, punctate. Thorax elongate, with the median line distinct towards the base. Elytra very densely punctate, thinly pubescent, and brown at the base and apex, with four subdistinct raised lines on each elytron. Apex of thighs, apex of tibiæ, apex of first joint of tarsi and all the others, brown. Abdomen dusky brown. Antennæ brown with the very base of each joint reddish.

ŒDEMERIDÆ.

696.—Selenopalpus fuscus. n. sp.

Length 5 lines.

Reddish brown, opaque, densely fulvo-pubescent. Head subtriangular, black. Neck large. Thorax elongate, and constricted in front, with the median line profoundly marked. Elytra elongate, subacuminate, coarsely punctate, and cinereo-pubescent, with the suture and extreme apex black. Body beneath and legs black, nitid and slightly cinereo-pubescent. Penultimate article of tarsi subbilobed. Antennæ short and slender, the joints subtriangular, setose, and of nearly equal length. Palpi serrate, last joint broadly triangular.

Neither this nor the following species answer exactly to the description given of the maxillary palpi of *Selenopalpus*, and as far as I can ascertain no other characters have been given to that genus. I have on this account been more particular in my description of the anatomy of the insects, than would have been necessary where the genus was properly defined.

697.—SELENOPALPUS MASTERSII. n. sp.

Length 3 lines.

Reddish-brown, opaque, punctate, hairy and cinereo-pubescent. Head broadly triangular, flat on the forehead and largely and roundly angled behind the eyes. Neck large and convex. Thorax black, subtransverse, subcordiform and covered with whitish hairs. Elytra coarsely punctate and densely cinereo-pubescent, the white pubescence scanty in some places, giving thereby the appearance of a large space near the scutellum, and a broad fascia behind the middle, not reaching the suture, of a reddish-brown colour. Under side of body, and legs piceous and subnitid. Penultimate joint of tarsi strongly bilobed. Antennæ red, longer and more slender than in the last species, and not setose. Maxillary palpi with the last joint of an elongate triangular form.

698.—ANANCA VITTICOLLIS. n. sp.

Length 4¼ lines.

Black, with the thorax and elytra pale red, the former oblong, a little rounded on the sides, and not broader behind than in front, with a large central black vitta not reaching the apex, Scutellum black. Elytra densely and minutely punctate, opaque, finely pubescent, and tricostate,

699.—ANANCA RUFICOLLIS. n. sp.

Length 3 lines.

Head black, thinly punctate. Thorax red, elongate, rounded

on the sides in front, and slightly narrowed behind. Elytra bluish black, densely and rugosely punctate. Meso- and meta-thorax, abdomen, and legs black, thighs red.

700.—PSEUDOLYCHUS APICALIS. n. sp.

Length 4 lines.

Black, opaque. Antennæ subpectinate. Head transverse. Eyes prominent. Thorax of the width of the head, transverse, and deeply impressed in the middle and sides. Elytra red with the apical third black, very densely punctate and marked with three or four slightly elevated longitudinal lines. Under surface subnitid.

The two following species were accidentally omitted in their proper places.

RHIPICERIDÆ.

701.—PSACUS MASTERSII. n. sp.

Length 2 lines.

Ovate, convex, black, subopaque, punctate, hairy, and marked on the sides of the thorax and elytra with white pubescence. Antennæ short, the flabellæ short and thick.

This species is smaller and more convex than *Psacus atta-genoides*, Pascoe, and looks even more like some of the family of *Dermestidæ* than that species.

DASCILLIDÆ.

702.—DASCILLUS BREVICORNIS.

Length 3½ lines.

Brownish red, subnitid, punctate, and densely clothed with a short yellow pubescence. Thorax transverse, of the width of the head at the apex, a little broader behind, and emarginate in the middle of the base to fit the scutellum. Elytra scarcely broader than the base of the thorax, and obsoletely striate, with the interstices subelevated. Legs and abdomen red. Antennæ shorter and less filiform than usual in the genus.

I subjoin a List of the Genera and Species described in this and my previous Paper.

Miscellanea Entomologica,

By William MacLeay, Esq., F.L.S.

[Read 7th July, 1873.]

In the month of January, 1870, I made a hurried entomological excursion into the Queanbeyan and Monaro districts. I was accompanied by Mr. Masters, of the Australian Museum, by whose assistance, notwithstanding the intense heat and dryness of the season, I was enabled to add a number of new and interesting species to my own as well as to the public collection.

Among the captures then made, there were two species of small Carabideous insects which peculiarly interested me. One of these has since been described by H. W. Bates, Esq., in the " Entomologists' Monthly Magazine" for July, 1871, under the name of *Eudalia Macleayi*. This genus has been associated, very properly, by Mr. Bates, with a group of insects named by Lacordaire *Anchonoderides*, and for which he proposes, and with good reason, the name of *Lachnophorinæ*.

I had previously, in the first volume of the Transactions of the Entomological Society of New South Wales, page 108, described a species from Port Denison as *Odacantha lutipennis*. Count Castelnau has since (Not. Aust. Col., 1867, page 16) described another species from Arnheim's Land, North Australia, under the specific name of *Waterhousei*. He also, at the same time, proposed for those insects the generic name of *Eudalia*, but without giving generic characters, and suggested their probable affinity to the subfamily *Ctenodactylides* of Lacordaire.

Mr. Bates, in the Paper previously referred to, not only describes the Monaro species, but gives the character of the genus.

The habit of the insect goes far to prove the correctness of Mr. Bates' hypothesis of its affinity to *Lachnophorus*. I found the species *E. Macleayi* abundant close to the water in the gravelly bed of the Umeralla River, a tributary of the Murrumbidgee, in

U

the **Monaro** district, and in similar positions in the Upper Murrumbidgee itself, and its tributary, the Queanbeyan River. How far its habitat extends upwards into the Snowy Mountains, or downwards along the course of these rivers towards the Plains, remains yet undetermined.

The other insect to which I referred, a small brilliant, brassy, black beetle, can scarcely be separated from the genus *Cymindis* of Latreille. It was found tolerably abundant under stones on the long sloping Downs beyond Cooma, and in the vicinity of Spring Flat. I subjoin a description of this, and another species which much resembles it.

CYMINDIS ÆNEA.

Long. 2¼ lin.

Æneo-nigra nitida, capite subconvexo, thorace subtiliter canaliculato, elytris viridi-nigris leviter striatis, corpore subtus femoribusque piceo-nigris, antennis palpis tibiis tarsisque piceo-rufis.

CYMINDIS ILLAWARRÆ.

Long. 3 lin.

Æneo-nigra nitida, thorace medio canaliculato ad latera marginato, elytris viridi-nigris leviter striatis, corpore subtus piceo-nigris antennis palpis pedibusque rufis.

This species, which, as its name indicates, was found in Illawarra, differs from *ænea* in being much larger, in having the legs entirely red, and in the form of the thorax, which is more regularly rounded, and more broadly margined on the sides.

The two species described by me (Trans. Ent. Soc. N. S. Wales, 1864, p. 111 and 112) under the names of *Cymindis longicollis* and *angusticollis*, do not belong to the genus, but probably belong to Baron de Chaudoir's genus *Xanthophœa*. *Cymindis curtula* Erichsen is a *Philophlaeus* and *Cymindis inquinata* of the same author is probably an *Agonocheila*. This reduces the number of Australian species of *Cymindis* to three, the two now described and *C. crassiceps*, described by me in a previous Paper.

The insect from the Clarence River, described below, has long been known to me, and I have often felt surprised that it should remain to the present time undescribed. I have repeatedly received specimens from the Clarence, and have seen them in other collections, so that it can scarcely be looked upon as a rare insect.

Genus LACHNODERMA.

Mentum subtransversum profunde emarginatum, dente medio magno valido obtuso lobis lateralibus longioribus apice subacutis.

Labium corneum subelongatum apice rotundatum.

Palpi lobiales validi securiformes.

Palpi maxillares validi oblique truncati.

Maxillæ apice arcuatæ acutæ.

Labrum quadratum antice ampliatum rotundatum.

Antennæ sublongæ filiformes articulo primo ceteris majori, secundo breviori.

Caput postice angustatum oculis prominentibus.

Thorax cordiformis transversus angulis posticis acutis recurvis.

Elytra thorace latiora subplana truncata.

Corpus alatum planum hirsutum.

Pedes subvalidi tarsorum articulo quarto fortiter bilobato.

LACHNODERMA CINCTUM.

Long. 5 lin.

Piceo-rufum subnitidum fortissime punctatum dense pilosum, thorace capite latiori lateribus antice valde rotundatis, elytris thorace latioribus cœruleo-cinctis, antennis pedibusque nigris.

The place of this insect will be with the *Helluonidæ*, a subfamily numerously represented in Australia, and of a well defined character.

Count Castelnau has given a very good review of the whole group in his "Notes on Australian Coleoptera." He divides them into those with wings and those without.

Those without wings he refers to the following genera:—

HELLUO Bon., to which only two species can now be referred—*costatus* Bon., from New South Wales, and *carinatus* Chaud., from Victoria.

PSEUDHELLUO Casteln., represented by one species, *Wilsoni* Casteln., an insect I have never seen.

ACROGENYS MacLeay, one species, *hirsuta* MacLeay, W.

HELLUODEMA Casteln., also represented by one species, *Batesii* Thomson.

The Helluonidæ with wings, Castelnau divides into the genera GIGADEMA Thomson, HELLUOSOMA Castelnau, and ÆNIGMA Newman. To these must now be added the present genus LACHNO-DERMA.

This winged division of the sub-family is very much more numerous than the unwinged, and I regret to find in Gemminger and Harold's "Catalogus Coleopterorum," that the three genera to which Count Castelnau referred them, have been merged into one—the *Ænigma* of Newman.

Mr. Masters has also made the same mistake in his Catalogue of Australian Coleoptera.

Gigadema and *Helluosoma* do not present any great differences of structure, still I think there are sufficient distinctions to justify Count Castelnau's classification, but as regards *Ænigma* there are wide and important differences which can leave no doubt as to the propriety of the subdivison. The species of these three genera hitherto described are :

GIGADEMA, 8 species—*grande* MacLeay, *Titanum* Thomson, *longipenne* Germar, *sulcatum* MacLeay, *politulum* MacLeay, *Bostockii* Casteln., *Paroense* Casteln., and *minutum* Casteln.

HELLUOSOMA, 5 species—*atrum* Casteln., *cyaneum* Casteln., *Mastersii* MacLeay, *resplendens* Casteln., and *cyaneipenne* Hope.

ÆNIGMA, 3 species—*Iris* Newman, *splendens* Casteln., and *Newmanni* Casteln.

Ænigma unicolor Hope, from Port Essington is believed by Count Castelnau to be a *Gigadema*.

The following are new species in my collection :—

HELLUOSOMA ATERRIMUM.

Long. 8 lin.

Nigrum subnitidum albo-hirtum, thorace cordato subquadrato punctato medio plano anguste canaliculato, elytris thorace latioribus planis punctatis subleviter striato-punctatis, antennis tarsisque fulvo-hirtis.

This species is from Cape York. It is easily distinguished from *H. atrum* Casteln, by its broader and flatter form, less elongate thorax, finely striated elytra, and hairy clothing; and from *H. cyaneum* and *cyanipenne*—the only other described species at all resembling it—by difference of colour, more elongate thorax and smoother elytra.

GIGADEMA DAMELII.

Long. 9½ lin.

Nigrum subopacum, thorace punctato cordato subquadrato, elytris octo-sulcatis sulcis biseriatim punctatis medio striatis, corpore subtus nitido.

This insect is also from Cape York, and is very like *G. sulcatum*. In fact the only important difference is to be found in the form of the thorax which in *sulcatum* is shorter and more broadly rounded than in the present species. The facial grooves near the eyes are more strongly marked also in this insect.

ÆNIGMA PARVULUM.

Long. 6 lin.

Cyaneum nitidum hirtum punctatum, capite subviridi antice nigro, thorace cordato transverso postice valde angustato medio leviter canaliculato, elytris punctatis striato-punctatis, corpore subtus subviridi glabro.

This insect is labelled "New Holland," a mark indicating simply that I have no record of the locality from which it came, nor how I got it. I am pretty sure, however, that it must have come from the Clarence or Richmond River districts. It is the smallest of the genus, and differs much from the other species in the form of the thorax, which is more like that of a *Helluosoma*. Indeed its resemblance generally to *Helluosoma resplendens* Casteln. is very great.

In page 96 of this volume of the Transactions of the Entomological Society of New South Wales, I gave the characters of a genus which I named *Philoscaphus* to a group of *Scaritidæ* up to that time included in the genus *Carenum*. Of that group I had previously described two species under the names of *Carenum tuberculatum* and *carinatum*.

The first of these is found over a large surface of country extending from East to West from the Murrumbidgee to South Australia, and probably having an equally wide range from North to South. It is not, however, an insect which can be called common anywhere, in that respect resembling almost all the Australian *Scaritidæ*.

The second species *Carenum carinatum* is very different from the first in size and appearance, and is one of the rarest of insects : I have only known of its being taken at Wingelo, near Goulburn, and at Bungendore, in the Lake George basin.

A third species *Philoscaphus Mastersii*, I described in the Paper named " The Insects of Gayndah " (ante page 96), and I mentioned then that I knew of two other species in collections. One of these from Nicol Bay I propose to name.

PHILOSCAPHUS COSTALIS.

Long. 14 lin.

Niger, thorace transverso lunulato capite latiori postice lobato, elytris ovatis scabris subtuberculatis granulatis sulcis lateralibus binis.

Except in having a double lateral groove on the elytra, this species is scarcely distinguishable from *tuberculatus* and *Mastersii*. This groove which lies between the costa which forms the apparent side of the elytra as seen from above, and the true lateral costa, and extends from the humeral angle to near the apex, is in this species divided by an intermediate costa, which takes its rise near the humeral angle and extends towards the apex as far as the others.

PHILOSCAPHUS LATERALIS.

Long. 11 lin.

Niger, thorace transverso lunulato capite latiori postice lobato, elytris ovatis ad basiu emarginatis obsolete quinqueseriatim tuberculatis sulcis lateralibus binis inæqualibus.

This species is from South Australia, and 1 believe unique. Like the last it has the lateral groove double, but the intermediate costa in the present insect takes its rise from the upper one at some distance from the humeral angle, and diverges gradually towards the apex, while the upper groove is throughout narrower than the lower.

Since my last Paper on the Australian *Scaritidæ*, a few species have come under my notice, which I shall now describe.

CARENUM PARVULUM.

Long. 5½ lin.

Nigrum subangustum, antennis gracilibus submoniliformibus, palpis labialibus apice triangularibus truncatis, thorace subquadrato postice rotundato, elytris oblongo-ovatis thorace angustioribus subopacis anguste violaceo-marginatis postice bipunctatis, tibiis anticis extus tridentatis.

This species was found by Mr. Masters and myself last summer near Murrurundi, a town situated on the upper valley of the Hunter, close to the Liverpool range.

It is remarkable in several respects. It is the smallest *Carenum* I have seen. The antennæ most resemble those of the *C. Spencei* group, the palpi are not securiform, and the anterior tibiæ are strongly tridentate. It would seem to form a kind of link between the groups of which *C. coruscum* and *C. Spencei* are the types.

CARENUM FOVEIPENNE.

Long. 7½ lin.

Nigrum, thorace subtransverso postice rotundato vix lobato anguste viridi-marginato, elytris oblongis antice truncatis postice rotundatis anguste viridi-marginatis quadri-seriatim foveatis foveis rotundis aureo-viridibus, tibiis anticis extus tridentatis dentibus validis subobtusis.

Mr. Odewahn of Gawler Town, South Australia, sent me this species. It differs chiefly from *C. Spencei* in the colouring of the margins, in having the rows of foveæ more distinctly separated by raised interstices, and by having a fiery green reflection at the bottom of each fovea.

CARENUM DIGGLESII.

Long. 7½ lin.

Nigrum subopacum, thorace subtransverso antice truncato postice lobato angulis posticis rotundatis subemarginatis, elytris ovatis antice emarginatis obsolete punctatis subcostatis triseriatim foveatis foveis magnis haud profundis, tibiis anticis extus tridentatis.

This species differs from *C. Spencei* and all the others of that group, in the sculpture of the elytra, which present three broad though faint costæ on each, with rows of almost obliterated punctures, and three rows of distant shallow foveæ, five or six in each row. The thorax also differs from that of *C. Spencei*, in having the posterior lobe more truncate, and the posterior angles more emarginate.

The only specimen of this insect I know belongs to Mr. Diggles of Brisbane, and was taken on a piece of timber floating down the Brisbane River during a flood.

CARENUM PLANIPENNE.

Long. 11 lin.

Nigrum nitidum, thorace transverso viridi-marginato lateraliter rotundato postice lobato emarginato angulis posticis emarginatis, elytris nitidissimis purpurascentibus viridi-marginatis ovatis subplanis postice bipunctatis, tibiis anticis extus bidentatis.

This is a South Australian species, and is found I believe at Port Wakefield. It is a broad, flat, beautiful insect, of the *C. marginatum* group, but differing in form and appearance very much from that insect. ·

CARENIDIUM LACUSTRE.

Long. 13 lin.

Nigrum nitidum cyaneo-marginatum, thorace subquadrato subdepresso postice angustato truncato, elytris elongato-ovatis punctis sublateralibus.

Like *C. Kreuslerœ* but of more elongate form. It is of a brilliant black, with the exception of a narrow margin of blue mingled with green on the thorax and elytra; the latter has a row of rather distant punctures along but not in the lateral margin.

My specimen was taken at Lake Albert, near Wagga Wagga, last September.

GNATHOXYS PUNCTIPENNIS.

Long. 5 lin.

Niger nitidus, thorace oblongo medio canaliculato, elytris subseriatim punctatis apice rugosis, antennis palpis pedibusque
  piceo-rufis, tibiis anticis extus bidentatis.

The smoothness of the sculpture in this species gives it very much the appearance of a *Promecoderus*. The puncturation on the elytra is not very marked, and the rugosity at the apex is very limited. The upper external tooth on the fore-tibiæ is small and near the middle of the joint.

My only specimen comes from South Australia.

The affinity of the insect described below is manifestly to *Oraspedophorus*.

I propose for the genus the name of

PLATYLYTRON.

Mentum transversum profunde emarginatum dente medio lato
  brevi lobis lateralibus rotundatis.

Labium corneum parvum truncatum apice bisetosum paraglossis
  adhærentibus superantibus.

Palpi maxillares longi articulo secundo longo, tertio brevi
  curvato, ultimo securiformi.

Palpi labiales valde securiformes.

Mandibulœ validæ apice acutæ arcuatæ intus dentatæ.

Labrum transversum leviter emarginatum.

Antennœ filiformes.

Caput parvum.
Thorax parvus lævis orbicularis.
Elytra magna ovata subplana.
Pedes ut in *Craspedophorus*.

PLATYLYTRON AMPLIPENNE.

Long. 9 lin.

Nigrum subnitidum, capite antice utrinque foveato, thoraco lævi antice subemarginato postice subtruncato ad latera rotundato late marginato—margine reflexo—dorso canaliculato ad basin utrinque profunde impresso, elytris subplanis latis oblongis profunde striatis—interstitiis convexis —lateribus apiceque rugose punctatis.

The head is smooth, and the fovea on each side of the face is short. The thorax is smooth and nearly circular, though a little emarginate in front and truncate behind, the dorsal line is deeply marked, and there are two deep impressions at the base near the posterior angles. The sides are rounded and not narrower behind than in front, with a broad reflexed margin. The elytra are broad, flat, rounded at the humeral angles, broadly rounded at the apex, deeply striated and roughly punctured towards the apex and sides. The legs are rather slender.

I received several specimens of this insect from King George's Sound some years ago.

———

Baron de Chaudoir has given the generic name of *Coptocarpus* to the insect formerly known as *Oodes Australis* Dej. He gives as the chief characters which distinguish it from *Oodes* proper—the adherence of the paraglossæ to the sides of the labium, the absence of spongy covering on a large portion of the under side of the first joint of the anterior tarsi, and the enormous dilatation of the three first joints of the same tarsi in the male. It is probable that these characters are common to all the *Oodes* of Australia, the tarsal characters certainly are in all those species which I have had an opportunity of examining. I shall therefore adopt Baron de Chaudoir's genus *Coptocarpus* for all the Australian *Oodes*, as I see has been done already by Mr. Masters in his catalogue of Australian Coleoptera.

Sixteen species have hitherto been described from various parts of this country, the majority of them from Western and Central Australia, but every portion of New Holland seems to be represented by two or more species. With the exception of *C. Australis* Dej. *fuscitarsis* Blanch., and *Reichei* Laferté, they have all been described by Count Castelnau in his " Notes on Australian Coleoptera " published in Melbourne in 1867.

My collection contains the following new species :—

COPTOCARPUS CHAUDOIRI.

Long. 5½ lin.

Niger subnitidus latus, epistomi sutura semicirculari bipunctata, thorace leviter canaliculato utrinque oblique subimpresso, elytris tenuiter striatis stria scutellari brevissima interstitiis planis, antennis palpis pedibusque rufis.

This species is from the Clarence River. It is of a broader form than *C. Australis*, and differs both from it and *C. fuscitarsis* in the extreme shortness of the scutellar stria, which is little more than an elongated point.

COPTOCARPUS RIVERINÆ.

Long. 7 lin.

Niger subnitidus, thorace tenuiter canaliculato haud lateraliter impresso, elytris leviter striatis stria scutellari scutello ter longiori, antennis rufis articulis primis et tertiis subfuscis.

In size and general appearance very like *C. oblongus* Casteln. It is however less opaque, and has the scutellar stria a little longer, and the hind tibiæ straighter than that insect.

The insect from which I take my description is from the Murrumbidgee, but I have two imperfect specimens in my collection labelled " Interior S. Aust." which I believe to be of the same species.

COPTOCARPUS NITIDUS.

Long. 5 lin.

Niger nitidus convexus, thorace tenuiter canaliculato ad latera oblique impresso sub-piceo, elytris leviter punctato-striatis, antennis palpis tarsisque rufis.

This pretty little species is very like *C. Convexus* Casteln. There is no scutellar stria, but the stria which runs parallel and close to the suture is interrupted near the scutellum. Its habitat is Cape York.

In his " Notes on Australian Coleoptera," published in Melbourne, in 1867, Count Castelnau has distributed the *Cnemacanthidæ* of Australia among six genera—*Mecodema* Blanch., *Maoria* Casteln., *Promocoderus* Dej., *Parroa* Casteln., *Adotela* Casteln., and *Cerotalis* Casteln. The two first of these are chiefly found in New Zealand, and are sufficiently distinct, but the other four genera are much alike, and in fact were all classed as *Promecoderus* until the publication of the Count's Paper. The new genera I believe to be undoubtedly good, but Count Castelnau has not done himself or his genera justice in the very careless way in which he has characterized them.

According to his statement, the distinctive characters may be summed up as follows :—

PROMECODERUS : Mentum with the lateral lobes round, and a slightly bifid median tooth. Palpi oval and truncate. Four first joints of the anterior tarsi dilated in the males, and furnished beneath with spongy brushes.

PARROA : Mentum large, without median tooth, and pointed at the inner angles of the lateral lobes. Palpi thick, oval, and truncate. Four first joints of the anterior tarsi in the males triangular, slightly dilated, and armed beneath with strong hairs.

ADOTELA : Mentum like *Parroa*. Palpi strongly securiform. Three first joints of anterior tarsi dilated, and furnished beneath with spongy brushes.

CEROTALIS : Mentum with median tooth, lateral lobes rounded.

Palpi as in *Promecoderus.* Four first joints of anterior tarsi in the males furnished beneath with spongy brushes or cushions.

It will be seen from this that the last named genus approaches very closely to *Promecoderus.*

I find, however, that the labium also differs somewhat in these genera. In *Promecoderus* it is subtruncate, but with a prominence in the middle, armed with two long setæ. In *Adotela* it is quite truncate, while in *Cerotalis* it is elongate and subemarginate. The labrum also in the two latter genera is more emarginate than in *Promecoderus.*

The insect described by me in the "Insects of Gayndah" (ante, page 99) as *Promecoderus viridis,* is undoubtedly an *Adotela.*

The following are new species :—

PROMECODERUS PARVULUS.

Long. 4½ lin.

Elongatus convexus fusco-niger nitidissimus, thorace ovato canaliculato, elytris angustis subparallelis haud striatis, corpore subtus piceo-nigro nitido, antennis palpis pedibus elytrorumque margine inflexo rufis.

This species is abundant enough in the Upper Murrumbidgee country, near Yass. It scarcely differs, except in size and colouring, from the South Australian species, *P. gracilis* of Germar.

PROMECODERUS RIVERINÆ.

Long. 6 lin,

Elongatus convexus niger nitidus, thorace elongato-ovato canaliculato, elytris angustis haud striatis, antennarum articulo primo palpis coxis tarsisque rufis.

As its name implies, this insect inhabits the Lower Murrumbidgee or Riverina country, and, like the last, closely resembles a South Australian species, *P. concolor* Germar. It may be readily distinguished from that species by its smaller size and more elongated thorax.

PROMECODERUS INTERRUPTUS.

Long. 6 lin.

Elongatus convexus niger nitidissimus, thorace subelongato

canaliculato, elytris ovatis striis dorsalibus interruptis, au-
tennarum articulo primo palpis pedibusque piceis.

This species is marked on the elytra near the suture with
what may be termed the semiobsolete interrupted traces of large
striæ. It is from the Clarence River, and is the only species I
have seen from that part of the country.

PROMECODERUS HUNTERIENSIS.

Long. 6 lin.

Brunneo-niger nitidissimus, thorace subelongato leviter canaliculato, elytris elongato-ovatis versus suturam striatis striis
subtiliter punctatis, corpore subtus nigro, antennis palpis
tarsisque piceo-rufis.

The punctato-striate elytra would place this insect in Castelnau's group composed of *P. maritimus, striato-punctatus, Bassii*
and *Wilcoxii*, whether the resemblance goes further I cannot say,
as I have never seen any of these four species. The two striæ
next the suture are very distinct, the next three or four are very
faint, and the sides are quite smooth. The punctures in the
striæ are all very minute. The colour is an olive or bronzy black
very nitid. Its habitat is the Hunter River district.

PROMECODERUS MASTERSII.

Long. 7 lin.

Æneo-brunneus nitidus, thorace suboblongo lateraliter rotundato canaliculato, elytris elongato-ovatis sub lente obsolete
punctatis striatis striis prope suturam magis profundis
lateribus a medio sulcatis, corpore subtus nigro, antennis
palpis tarsisque piceo-rufis.

The median line of the thorax is very well marked, the two
striæ on each side of the suture of the elytra are also well marked,
and the sublateral groove extends from the middle or almost
before it to the apex, where it is widened and biimpressed. These
characters distinguish this species from *P. brunnicornis* Dejean.

Its habitat is Monaro.

PROMECODERUS INORNATUS.

Long. 6 lin.

Niger subnitidus, thorace subelongato tenuiter canaliculato, elytris elongato-ovatis striatis postice sulco sublaterali brevi tripunctato, segmentis abdominalibus 3, 4, et 5, utrinqne fortiter impressis, antennis palpis tarsisque piceo-rufis.

This is also a Monaro insect. The head has a large transverse impression behind the eyes, and two impressed punctures on each side in front, the posterior being on the suture of the epistome. The median line of the thorax is very lightly marked. The elytra are distinctly striated throughout though more faintly towards the sides, while under a strong lens the striæ appear green and indistinctly punctate. The sublateral groove is much shorter than in *P. Mastersii*, and seems to take its rise from a large puncture. The 3rd, 4th, and 5th abdominal segments have on each side a very deep round impression, but this is more or less common to every species of the genus. The whole insect is of a rather dullish black hue.

PROMECODERUS PUNCTICOLLIS.

Long. 5 lin.

Niger subnitidus, capite antice utrinque leviter bipunctato, thorace subelongato postice subrotundato medio tenuiter canaliculato disco ante medium utrinque bipunctato punctis oblique positis, elytris fortiter striatis striis ad latera subobsoletis subpunctatis, antennarum articulo basali rufo.

This species, also from Monaro, is deeply striated on the elytra, excepting close to the sides. The thorax is rather long, is rounded at the base, has the median line indistinctly marked, and has two diagonally placed punctures near each side in advance of the middle. In other respects it resembles *P. inornatus*.

PROMECODERUS DORSALIS.

Long. 5 lin.

Brunneo-niger subæneus nitidus, capite postice nigro-viridi, thorace subelongato fortiter canaliculato antice triangu-

lariter transversim impresso, elytris striatis striis tribus
versus basin distinctis ceteris subobsoletis, corpore subtus
nigro segmentis abdominalibus leviter impressis, antennis
palpis coxis tarsisque piceo-rufis.

A common species on the Murrumbidgee. It is small and
narrow. The back of the head is greenish. The thorax has the
median line deeply impressed and terminating some distance
from the apex in a deep emargination of the transverse impression. The three striæ nearest the suture on the elytra are
tolerably well marked on the basal half, the others are very
faint.

PROMECODERUS ANTHRACINUS.

Long. $6\frac{1}{2}$ lin.

Niger nitidissimus, capite inter oculos utrinque leviter canaliculato postice transversim impresso, elytris ad suturam
bistriatis striis apicem haud attingentibus, segmentis
abdominalibus utrinque foveatis haud transversim striatis,
antennarum articulo primo rufo.

From the Lower Murrumbidgee. It is of elongate form.
The head is shallowly impressed on each side between the eyes.
The thorax is lightly marked on the median line, and has a transverse impression near the base. The striæ on the elytra are
almost obsolete, excepting two on each side of the suture, which
are tolerably well marked for two-thirds of their lengths from
the base. The sublateral groove is very short. All the abdominal segments have a deep fovea on each side, but without transverse striæ.

PROMECODERUS OLIVACEUS.

Long. 6 lin.

Æneo-brunneus nitidus, capite nigro, thorace subelongato
canaliculato, elytris ovatis striatis ad latera glabris sulco
sublaterali brevi, corpore subtus nigro, segmentis abdominalibus utrinque leviter foveatis et canaliculatis, antennis
palpis coxis tarsisque piceo-rufis.

This insect was taken last summer at Piper's Flats, on the western side of the Blue Mountains. It is of a stout convex form, and differs from *P. brunnicornis* not only in this respect, but in being smaller, and in having the median line of the thorax and the striæ of the elytra more distinctly marked.

ADOTELA NIGERRIMA.

Long. 7 lin.

Convexa nigra laevis nitidissima, thorace subquadrato canaliculato ad latera rotundato postice subangustato, elytris ovatis apice subacuminatis subrugosis lateribus postice quadripunctatis vix sulcatis, tibiis anticis gracilibus, antennarum articulis sex ultimis pubescentibus.

My only specimen of this insect is a female, and, consequently, the remarkable form of the palpi and tarsi does not appear, but I have no doubt of its belonging to this genus.

It is from the Percy Islands.

———

Among a number of new and very interesting Coleoptera brought by Mr. Damel from Cape York some years ago, was an unique specimen of a very remarkable and gigantic Carabideous insect. It has in general appearance a considerable resemblance to an insect found in the deserts of the Caspian, *Dioctes Lehmanni* Menetries, about the affinities of which entomologists have widely disagreed.

Baron de Chaudoir believes its affinity to be with *Acinopus*, while Lacordaire places it among the *Cnemacanthidæ*.

The difficulty felt in the case of the genus *Dioctes* will, I suspect, extend to the present insect, though, beyond the general resemblance, there is not much in common in the two insects.

The following are the characters, as nearly as I can give them from the rather mutilated specimen in my possession, of this genus which I propose to name

MECYNOGNATHUS.

Mentum magnum profunde emarginatum dente medio brevi lato bifido lobis lateralibus rotundatis intus verticalibus.

v

Labium longum subangustum truncatum crassum longitudinaliter carinatum apice bisetosum paraglossis liberis palpi-
formibus vix labium superantibus.

Maxillæ subelongatæ subgraciles.

Palpi maxillares longi apice oblique truncati articulo secundo
longissimo quarto penultimo breviori.

Palpi labiales longi apice subtriangulares oblique truncati
articulo secundo longissimo.

Mandibulæ maximæ porrectæ intus lobatæ.

Labrum abest.

Antennæ sublongæ filiformes articulo secundo breviori.

Caput maximum crassum subquadratum clypeo profunde emarginato ; subtus concavum utrinque fortiter carinatum.

Thorax brevis subangustus postice angustatus.

Elytra ovalia subconvexa.

Pedes sublongi haud validi tibiis inermibus tarsis subtriangulari-
bus.

Corpus apterum pedunculatum.

MECYNOGNATHUS DAMELII.

Long. 18 lin. (mand. incl.)

Niger subopacus, capite plano antice subcirculariter impresso
angulis posticis carinatis, thorace marginato capite angustiori subcordato medio leviter canaliculato postice transversim impresso, elytris ovalibus obsolete striatis marginatis,
pedibus piceis.

The most striking peculiarity about this insect is the enormous
size of the head and mandibles, the two together nearly equalling
in length the remainder of the body.

The right mandible which is broken off at about half its
length, has a pointed lobe on its inner surface ; the left has a
more rounded lobe on its inner surface about the middle, and
terminates in an obtuse point without the slightest tendency to
curve inwards.

The maxillary palpi are as long as the mandibles, and have
as well as the labial palpi the second joint much longer than the

others. The antennæ have the first joint thicker, but not longer than the rest; the second shorter, and from the fourth they become evidently more attenuated. The labium is wanting, but the edge of the epistome is deeply emarginate. The head above is nearly square, flat on the vertex, and hollowed out in front, with a lateral carinuation, which is curved behind so as to form the posterior angles; beneath it is concave and canaliculate in the middle, while near each side there is a strong rugose ridge which terminates suddenly near the base.

The form of the thorax is also very unusual, it is flat, broadest in front, but scarcely so broad as the head, narrowed rapidly to the base and subtruncate in front and behind, with a lateral re-curved margin broadest at the posterior angles. The elytra are of an almost perfectly oval form. The legs are not strong for the size of the insect; indeed, the tibiæ are slight and without external teeth, the tarsi are also rather slight, the first joint is long, the rest get gradually shorter.

I can scarcely venture to give an opinion as to the position of this remarkable insect among the *Carabidæ*.

It would appear quite as much out of place among the *Cnemacanthidæ* or *Stomidæ* as *Dioctes*, and I cannot see any affinity in either genus to *Acinopus*.

On the other hand, there is, judging by description, for I have never seen the genus myself, a marked approach in the anatomy of the mentum labium, &c., to *Luperca* of Castelnau.

———

All the Paussidæ of Australia belong to the genus *Cerapterus* of Swederus, a name applied to those of the family which have ten joints to the antennæ.

The genus has since been largely subdivided, and two of the subgenera *Arthropterus* MacLeay and *Phymatopterus* Westwood are peculiar to this country. Of the first of these there are many species though all extremely rare. *A. Macleayi* Don. the original species, long remained the sole representative of the group, to this Professor Westwood added about twenty years ago the species *brevis, denudatus, parallelocerus, subsulcatus*, and *Wilsoni*, and about a year ago I described in the "Insects of Gayndah" five others, *angusticornis, elongatulus, Kingii, Mastersii,* and *Westwoodii.*

In the present Paper I add largely to the number of species, but I regret to say that I cannot contribute much to our knowledge of the habits and history of these very curious insects. Two observations only can I add to the scanty information we at present possess, one is, that some of the species are nocturnal, as I have frequently known them to fly into lighted rooms; the other, that one species at least *A. brevis* Westw. is gregarious. On one occasion I found upwards of fifty of them clustered together under a small piece of the loose bark of a gum tree, and Mr. Masters informs me that he has met with as many as twenty or thirty in a similar situation. This by no means proves however that it is a bark insect, as in both the instances alluded to, the ground was very wet, and the insects may have taken to the tree for protection and shelter. I have repeatedly during floods in the Murrumbidgee taken large numbers of species of *Catadromus, Aptinus, Steropus, Poecilus,* and other undoubted ground beetles under the bark of trees.

In order to simplify to some extent what in such a homogeneous group must always be difficult—the detection of the species—the following synopsis will be useful:

ARTHROPTERUS.

Section 1.—Antennæ with the joints 2 to 9 three times or more broader than the length.

1st. Thorax longer than broad, body naked.

A. *Mastersii,* MacLeay, W.	A. *Macleayi,* Don.
—. *cylindricollis,* n. sp.	—. *elongatulus,* MacLeay, W.
—. *Waterhousei,* n. sp.	—. *bisinuatus,* n. sp.
	—. *angulicornis,* n. sp.

2nd. Thorax longer than broad, body hairy.

A. *Hopei,* Westw.	A. *montanus,* n. sp.
—. *Riverinæ,* n. sp.	—. *subampliatus,* n. sp.
—. *nigricornis,* n. sp.	—. *humeralis,* n. sp.
—. *picipes,* n. sp.	—. *ovicollis,* n. sp.

3rd. Thorax as broad as long.

A. brevis, Westw.		*A. foveicollis*,	n. sp.
—. *parallelocerus*, Westw.		—. *Odewahnii*,	n. sp.
—. *Westwoodii*, MacLeay, W.		—. *scutellaris*,	n. sp.
—. *Wyanamattæ*,	n. sp.	—. *Turneri*,	n. sp.
—. *angulatus*,	n. sp.	—. *hirtus*,	n. sp.
—. *subcylindricus*,	n. sp.	—. *Darlingensis*,	n. sp.
—. *Adelaidæ*,	n. sp.	—. *depressus*,	n. sp.
—. *puncticollis*,	n. sp.	—. *Rockhamptonensis*,	n. sp.

Section 2.—Antennæ with the joints 2 to 9 less than three times broader than the length.

A. Wilsoni, Westw.	*A. politus*, n. sp.
—. *Howittii*, n. sp.	—. *latipennis*, n. sp.
—. *brevicollis*, n. sp.	—. *angusticornis*, MacLeay, W.
—. *denudatus*, Westw.	—. *Kingii*, MacLeay, W.

The first of these groups contains two distinct types. *A. Mastersii cylindricollis* and *Waterhousei* constitute one of them. They are insects of large size, with the antennæ moderately broad, the thorax elongate and widest in the middle, and the elytra long and truncate. The others are of the *A. Macleayi* type—antennæ short and broad, and thorax almost parallel-sided.

The second group, of which *A. Hopei* Westw.—an insect placed most unaccountably in Gemminger and Harold's Catalogue in the genus *Phymatopterus*—may be taken as the type, has the antennæ broad, the thorax broadest in the middle, and the elytra more or less covered with long hair.

The third group is the most numerous in species, and the most difficult of definition of the whole genus. The antennæ are very short and broad, the thorax is more or less quadrate, broadest in front, and rounded and ciliated at the anterior angles, the tibiæ are of great width, and the elytra are for the most part short, with a corneous apex. In explanation of this last peculiarity, I may state that all the *Cerapteri* have the extreme edge of the termination of the elytra thinned down as it were into a corneous semitransparent substance, and in most species this corneous edge is more or less emarginate and sinuate. In the present group this is particularly the case. I have long had an idea that the great variety of outline exhibited in this way would be found to give good specific characters, but as I have not found any one to make accurate sketches of the insects, I have been unable to test the value of my theory.

A. brevis is easily known from the others by its breadth, brilliancy, and subcordiform thorax, but *Westwoodii, parallelocerus, Wyanamattæ, angulatus, subcylindricus, Adelaidæ, puncticollis,* and *Odewahnii* come so near one another that it is most difficult to find good distinctive characters. I am satisfied, however, that I have not made too many species ; I believe, in fact, that I have three or four more species of this type in my collection. *A. foveicollis* has very distinctive marks, *scutellaris* and *Turneri* have a different form of thorax from the others, *hirtus* and *Darlingensis* have the elytra hairy, and *depressus* and *Rockhamptonensis* lead off towards the next group.

Section 2 consists of one group only. It comprises all the species with narrow antennæ and tibiæ. Commencing with *A. Wilsoni,* in which the thorax is rather broader than long, it terminates with *A. Kingii,* in which the thorax and general form exhibits a near approach to the insects of the first group—*Mastersii, cylindricollis,* and *Waterhousei.*

The species described by Professor Westwood under the name of *subsulcatus,* from King George's Sound, is unknown to me, and I cannot tell from his description to which of the above groups it would belong.

I subjoin descriptions of the new species.

ARTHROPTERUS CYLINDRICOLLIS.

Long. 6 lin., lat. $1\frac{1}{2}$ lin.

Piceus nitidus, antennis latis articulo primo quadrato nitido, secundo ceteris angustiori, capite inter oculos bi-impresso, thorace capite angustiori elongato subcylindrico rugoso punctato ad latera antice subrotundato postice subconstricto ad basin truncato linea dorsali leviter impressa, elytris elongatis parallelis tenuiter punctatis subpilosis pilis brevibus fulvis, tibiis apice acute angulatis.

Hab. Rockhampton.

Of a narrow elongate form and pitchy colour, sparsely covered with a short light red pile. The antennæ are broad, the first joint is square and rectangular, the second is narrower than the third, the rest are equal in width, but the last is as long as three others put together. The head is abruptly angled behind, forming a very marked neck. The thorax is narrower than the head and much longer than broad, is coarsely and somewhat rugosely punctured, has the sides a little rounded before the middle and constricted near the base, and has the median line only visible in the somewhat flattened middle part of the back. The external angles of all the tibiæ are acute.

ARTHROPTERUS WATERHOUSEI.

Long. 6 lin., lat. 1¾ lin.

Picco-rufus nitidus, capite inter oculos depresso, antennis sublatis articulo primo subtransverso, thorace subelongato antice angustato dorso punctato canaliculato transversim striolato, elytris thorace latioribus parallelis truncatis subtiliter punctatis punctis setigeris, tibiis posticis extus obtuse terminatis.

Hab. South Australia.

I name this species after Mr. Waterhouse of South Australia, to whom I am indebted for many valuable insects from that territory. It in some respects much resembles *A. Mastersii*, but is of narrower form and redder colour, has the antennæ less broad and the thorax coarsely punctured and depressed on the median line.

ARTHROPTERUS BISINUATUS.

Long. 3½ lin.

Picco-rufus subnitidus punctatus, antennis latissimis articulo primo transverso, 2, 3, 4 gradatim latioribus 5, 6, 7, 8, 9 longitudine quinquies latioribus, thorace oblongo subparallelo dorso subdepresso medio subtilissime canaliculato, clytris apice conjunctim emarginatis, tibiis anticis extus acute intermediis et posticis obtuse terminatis.

Hab. Lane Cove, near Sydney.

This species resembles *A. Macleayi*, but may be readily distinguished from it "inter alia," by the gradually increasing width of the antennæ and their bisinuate joints.

I have never seen but one specimen of this insect, and that was taken many years near Lane Cove by that most excellent collector the late Mr. Turner.

ARTHROPTERUS ANGULICORNIS.

Long. $4\frac{1}{2}$ lin., lat. $1\frac{1}{2}$ lin.

Rufo-piceus subnitidus crasse punctatus, antennis latissimis articulo primo transverso angulo externo subproducto, thorace elongato dorso subdepresso lateribus subparallelis, elytris postice leviter punctatis subtruncatis, tibiis posterioribus extus obtuse terminatis.

Hab. Ipswich, Moreton Bay.

This also approaches *A. Macleayi*. It is however of a darker colour and denser and coarser puncturation, has the exterior angle of the first joint of the antennæ produced, and has scarcely any trace of the median line on the thorax.

ARTHROPTERUS RIVERINÆ.

Long. $4\frac{1}{2}$ lin, lat. $1\frac{3}{4}$ lin.

Piceo-rufus nitidus fulvo-hirtus, antennis latissimis articulo primo transverso angulo externo subproducto, capite post oculos rotundato, thorace oblongo lateribus antice subrotundatis dorso subdepresso canaliculato, elytris leviter rugoso-punctatis apice singulatim rotundato-acuminatis, tibiis anticis extus acute intermediis et posticis obtuse terminatis.

Hab. Lower Murrumbidgee.

The thorax in this species is of very different form to that of *A. Hopei* and of the group generally. It is rather broad, not narrowed in front, and nearly parallel-sided. It differs also from *Hopei* in the more angular basal joint of the antennæ, the flatter and more deeply canaliculate thorax, and the more pointedly rounded apex of each elytron.

ARTHROPTERUS NIGRICORNIS.

Long. 5 lin., lat. 1¾ lin.

Piceo-rufus nitidus fulvo-hirtus, antennis latissimis piceo-nigris, thorace subelongato antice angustato dorso crasse punctato breviter canaliculato, elytris punctatis hirtis subtruncatis, pedibus piceo-nigris tibiis posticis obtuse terminatis.

Hab. Wide Bay.

The antennæ in this species are broad black and nearly paralleled-sided, the second and third joints being perhaps a shade wider than the others. The head is rounded behind the eyes. The thorax is oblong, coarsely punctured on the disc, rounded on the sides near the front, and canaliculate in the centre of the median line. The elytra are rather broad and depressed, punctate, fulvo-pilose, as long as the body, and subtruncate or very broadly rounded at the apex.

ARTHROPTERUS PICIPES.

Long. 4½ lin., lat. 1½ lin.

Rufo-piceus nitidus fulvo-hirtus, antennis latissimis, capite punctato inter oculos utrinque rectangulariter impresso angulis posticis rotundatis, thorace oblongo antice angustato ad latera subrotundato dorso crasse punctato profunde canaliculato, elytris punctatis hirtis apice singulatim subrotundatis, pedibus piceis, tibiis anticis extus acutissime terminatis.

Hab. South Country, near Yass.

Very like *A. Hopei.* It is of a darker colour, has the antennæ broader and longer, and the thorax more deeply canaliculate.

ARTHROPTERUS MONTANUS.

Long. 5 lin., lat. 1½ lin.

Piceus nitidus fulvo-hirtus, antennis latissimis articulo primo parvo quadrato rectangulari, thorace oblongo punctato profunde breviter canaliculato antice angustato postice sub-

constricto lateribus medio subrotundatis, elytris leviter
punctatis apice subtruncatis, tibiis posterioribus extus ob-
tusissime terminatis.

Hab. Bombala, Australian Alps.

Of narrower form, and much darker colour than *A. Hopei.*
The basal joint of the antennæ is small and square. The eyes
are prominent, and black. The thorax is narrowed in front,
rounded before the middle, constricted near the base, and deeply
canaliculate on the basal two-thirds of the median line.

ARTHROPTERUS SUBAMPLIATUS.

Long. $4\frac{1}{2}$ lin., lat. $1\frac{3}{4}$ lin.

Rufo-piceus subnitidus fulvo-hirtus, antennis latissimis articulo
primo transverso angulato, thorace oblongo parce punctato
canaliculato antice vix angustato, elytris postice subampliatis
apice subtruncatis, tibiis anticis extus acute posterioribus
obtuse terminatis.

Hab. Bombala.

This is a shorter, broader, and less nitid insect than the last
from the same locality. The first joint of the antennæ is more
angular and transverse, the thorax is more depressed, less narrowed
in front and not constricted behind, and the elytra are wider and
somewhat ampliated towards the apex.

ARTHROPTERUS HUMERALIS.

Long. $5\frac{1}{2}$ lin., lat. 2 lin.

Rufo-piceus subnitidus fulvo-hirtus, antennis latissimis articulo
primo transverso, thorace subelongato subdepresso punctato
medio leviter canaliculato antice angustato postice truncato,
elytris longis parallelis subdepressis apice late rotundatis
conjunctim subemarginatis, tubiis anticis extus subacute
terminatis.

Hab. Wellington and Dabee.

This is a large, dull coloured, and very hairy insect. The
thorax is less deeply canaliculate than in *A. Hopei.* The elytra
are prominently rounded at the humeral angles, and broadly

rounded and conjointly emarginate at the apex. The fore tibiæ are less acutely terminated at the external angle than in the other species of the group. The resemblance between this species and *A. picipes* is very great.

ARTHROPTERUS OVICOLLIS.

Long. 6½ lin., lat. 2 lin.

Rufo-piceus subnitidus subhirtus, antennis sublatis articulo primo quadrato, capite post oculos obtuse producto, thorace subovato postice transversim impresso dorso punctato profunde canaliculato, elytris leviter punctatis apice late rotundatis, tibiis haud latissimis.

Hab. South Australia.

The comparatively narrow antennæ and tibiæ and short hair of this species, are sufficient of themselves to distinguish it from all the others of the *Hopei* group, while the width and small elongation of the thorax are additional proofs of its aberrant character. It is in fact a near approach to the insects of the second section—*A. Wilsoni* and others.

ARTHROPTERUS WYANAMATTÆ.

Long. 4½ lin., lat. 1½ lin.

Piceo-rufus nitidus crebre punctatus, antennis latissimis articulis 2-9 brevissimis, oculis pallidis, thorace quadrato postice subangustato dorso depresso canaliculato ad latera antice rotundato ciliato, scutello transversim profunde impresso, elytris apice sinuato-truncatis, tibiis posticis extus apice oblique truncatis.

Hab. Camden.

Two species nearly resembling this insect have been described *A. parallelocerous*, Weston, and *Westwoodii*, MacLeay, W. The first of these, or rather what I take to be it, is a Melbourne insect, and is much less densely punctured and more nitid than *A. Wyanamatta*, the other from Gayndah is of a darker colour, has the basal joint of the antennæ more produced at the external angle, has the thorax more narrowed at the base and less densely

punctured, and has the elytra more hairy. The transverse depression on the scutellum is very deep in the present species. I
have some specimens in my collection from the same locality as
this insect, which differ in having black eyes and being of a
darker hue, but I scarcely think they can be of a distinct species.

The name *Wyanamatta* is derived from the Geological term
for the shale, which abounds in the district of Camden.

ARTHROPTERUS ANGULATUS.

Long. 4½ lin., lat. 1½ lin.

Piceus nitidus crebre punctatus, antennis latissimis brevissimis
articulo primo lato angulato, thorace quadrato plano canaliculato ad latera antice late rotundato postice subangustato
ciliato, elytris subtruncatis angulis humeralibus ciliatis, tibiis
posterioribus apice late rotundatis.

Hab. Rockhampton.

This species is of a very dark colour, densely punctured, and
covered on the sides of the thorax and elytra with short strong
hairs. The first joint of the antennæ is broad and produced at
the external angle. The thorax is broad, flat and very broadly
rounded at the anterior angles.

ARTHROPTERUS SUBCYLINDRICUS.

Long. 5 lin., lat. 1½ lin.

Piceo-rufus nitidus punctatus, antennis latissimis brevissimis
articulo primo subacute angulato, thorace capite latiori
quadrato subplano canaliculato angulis anticis rotundatis,
scutello ad basin transversim impresso, elytris apice leviter
trisinuatis, tibiis posticis extus apice oblique truncatis.

Hab. Bogaloug, near Yass.

I believe this species to have a wide range throughout the inland districts of New South Wales. It is more nitid, less punctate, and less hairy than the last, and is also of a lighter hue.
The thorax though much rounded near the anterior angles, has a
very broad square aspect. The transverse impression on the
scutellum is near the base, and not very deep.

ARTHROPTERUS ADELAIDÆ.

Long. 4 lin., lat. 1¼ lin.

Piceo-rufus nitidus parce punctatus, antennis latissimis brevissimis articulo primo transverso angulo externo subproducto, thorace subquadrato canaliculato ad latera antice subrotundato postice leviter angustato obsolete transversim impresso, elytris apice subtruncatis minute tri-emarginatis, tibiis posterioribus apice extus obtuse angulatis.

Hab. South Australia.

The thorax is somewhat narrower, and more of the *A. Macleayi* type than the preceding three species. It is also smaller, of a lighter colour, and more sparsely punctate.

ARTHROPTERUS PUNCTICOLLIS.

Long. 5½ lin., lat. 2 lin.

Piceus subnitidus rude punctatus, antennis latissimis brevissimis nigris articulo primo transverso angulo externo vix producto, oculis nigris, thorace quadrato ciliato dorso obsolete punctato profunde canaliculato—canali antice posticeque abbreviato—ad latera antice rotundato postice angustato, elytris latis ciliatis apice tri-emarginatis, tibiis posticis apice extus subtruncatis.

Hab. Liverpool Plains.

This is a large, dark coloured, coarsely punctured species. The thorax is deeply channeled in the middle, but the depression does not extend to the apex or base. On each side of the median line, in advance of the middle, there is a large indistinct puncture ; this, however, may not be constant, and I have never seen more than one individual of the species.

There is a marked resemblance to *A. subcylindricus.*

ARTHROPTERUS FOVEICOLLIS.

Long. 5 lin., lat. 1½ lin.

Rufo-piceus subnitidus punctatus, capite inter oculos minute bi-impresso, antennis latissimis brevissimis articulo primo subquadrato, thorace subquadrato dorso postice canaliculato antice bifoveato angulis anticis rotundatis posticis rectis subfoveatis, clytris subsinuato-truncatis, tibiis apice extus subacutis.

Hab. near Sydney.

This is a very distinct species, and apparently very rare.

The colour is dark, the head is marked between the eyes by two small depressions, the thorax has the median line well marked near the base, and has a deep fovea on each side of it near the front.

ARTHROPTERUS ODEWAHNII.

Long. 5 lin., lat. 1¼ lin.

Rufo-piceus subnitidus punctatus subangustus, antennis latissimis brevissimis articulo primo transverso 2-5 gradatim latioribus, oculis nigris, thorace subquadrato dorso postice canaliculato ad latera antice subrotundato postice leviter angustato, elytris emarginato-truncatis, tibiis posterioribus extus obtuse terminatis.

Hab. South Australia.

This insect differs from *A. Adelaidæ*, the only other South Australian species in the group, in the more elongate form, the prominent black eyes, and narrower and less square thorax.

I received my specimen from Mr. Odewahn, of Gawler Town, after whom I have named it.

ARTHROPTERUS SCUTELLARIS.

Long. 4½ lin., lat. 1½ lin.

Piceo-rufus nitidus dense punctatus, antennis latissimis brevissimis articulo primo magno transverso angulato, oculis nigris, thorace subquadrato leviter canaliculato ad latera antice rotundato postice modice angustato angulis posticis rectis subrecurvis, scutello late impresso medio punctato, clytris apice subemarginato-rotundatis, tibiis posticis extus apice oblique truncatis.

Hab. South Country.

I have no record of the exact locality in which this insect was captured, but I think it comes from the neighbourhood of Yass.

The most noticeable feature about it, is the acute slightly re-curved posterior angle of the thorax. The punctured depression on the scutellum also is unusually large.

ARTHROPTERUS TURNERI.

Long. 4½ lin., lat. 1½ lin.

Piceo-rufus nitidus dense punctatus, antennis latissimis brevis-simis articulo primo parvo subtransverso, thorace subquad-rato postice canaliculato truncato ad latera antice rotundato angulis posticis rectis subrecurvis subfoveatis, scutello semicirculariter impresso, elytris singulatim subrotundatis, tibiis posterioribus apice subtruncatis angulis externis sub-acutis.

Hab. Lane Cove, near Sydney.

I name this insect after the captor, the late Mr. Turner of Lane Cove. It departs considerably in the rather elongate shape of the thorax, from the majority of the group in which I place it. The sides are well rounded before the middle, and somewhat narrowed near the base which is truncate, with the angles, sharp, slightly recurved, and impressed.

ARTHROPTERUS HIRTUS.

Long. 4 lin., lat. 1¼ lin.

Piceo-rufus subnitidus punctatus, antennis latissimis brevissimis articulo primo parvo subquadrato, oculis nigris, thorace quadrato subplano leviter canaliculato postice subangustato obsolete transversim impresso angulis anticis late rotunda-tis, elytris dense fulvo-hirtis subtruncatis; tibiis posteriori-bus apice extus obtuse terminatis.

Hab. Monaro.

The thick clothing of soft reddish hair on the elytra is the most marked charcteristic of this insect.

ARTHROPTERUS DARLINGENSIS.

Long. 5½ lin., lat. 1¾ lin.

Piceo-rufus subnitidus punctatus, antennis latis articulo primo parvo subquadrato, capite inter oculos triangulariter leviter impresso, thorace subquadrato medio breviter depresso ad latera antice rotundat opostice obsolete transversim impresso, elytris fulvo-hirtis apice singulatim subrotundatis, tibiis posterioribus extus obtuse terminatis.

Hab. Darling River.

This insect, perhaps, should be placed in the *A. Hopei* group, the fact of the thorax being scarcely longer than broad, is my only excuse for its present position. The thorax is peculiar. It has in the centre of the median line a short broadish depression, which seems to terminate at both ends in an ill-defined roundish fovea.

ARTHROPTERUS DEPRESSUS.

Long. 5 lin., lat. 2 lin.

Rufo-piceus subnitidus latus subplanus dense punctatus pilis brevibus vestitus, antennis latis sublongis, thorace subquadrato breviter canaliculato postice subangustato ángulis anticis rotundatis, elytris longis truncatis, tibiis extus sub-obtuse terminatis.

Hab. Tweed River.

This species is unlike any other. It is of a dark dullish red colour, densely punctate, and thickly clothed with a short fulvous pile. The antennæ are broad, but owing to the length of the joints, the width of each joint is not more than three times the length. The thorax is almost broader than the length, nearly truncate in front, and quite truncate behind, with the anterior angles rounded and the posterior square, slightly narrowed towards the base, and shortly canaliculate on the posterior half of the median line. The fore tibiæ are rather obtusely terminated at the external angle, the others still more obtusely.

ARTHROPTERUS ROCKHAMPTONENSIS.

Long. 5 lin., lat. 1¾ lin.

Rufo-piceus nitidus punctatus, antennis sublatis articulo primo parvo quadrato, capite inter oculos leviter bi-impresso, thorace subquadrato canaliculato ad latera antice rotundato postice subangustato angulis posticis rectis subrecurvis, elytris longis truncatis, tibiis sublatis posterioribus extus apice subtruncatis.

Hab. Rockhampton.

This insect approaches the species of the 2nd section, and indeed much resembles some of them, though the antennæ and tibiæ are not so very narrow as in that group. The specimen before me is, I believe unique, in the collection of Mr. Masters.

ARTHROPTERUS HOWITTII.

Long. 5 lin., lat. 1½ lin.

Rufus nitidus punctatus, antennis subangustis articulo primo parvo quadrato, 2-9 longitudine haud ter latioribus, oculis nigris, thorace subquadrato tenuiter canaliculato ad latera antice subrotundato postice vix angustato, elytris apice subrotundatis, tibiis subangustis extus acute terminatis.

Hab. Victoria.

The thorax in this species is very much of the same character as in *Westwoodii*, *Wyanamattæ*, &c., but the antennæ and tibiæ are very different. The joints 2 to 9 of the former are not three times broader than long but are more than twice. The tibiæ are wider than in *A. Wilsonii* but very much narrower than in the preceding group.

ARTHROPTERUS BREVICOLLIS.

Long. 4½ lin., lat. 1½ lin.

Piceo-rufus nitidus leviter punctatus, antennis subangustis articulo primo parvo quadrato ceteris longitudine bis latioribus, oculis pallidis, capite inter oculos depresso, thorace subtransverso tenuiter canaliculato postice transversim impresso, elytris longis apice subemarginato-rotundatis, tibiis subangustis extus acuto terminatis.

Hab. New South Wales.

Very like the last species, but differs in the narrower antennæ and thinner puncturation and in having the thorax a little broader than the length. It approaches *A. denudatus* Westw. or at least what I take for that insect, for I am not by any means sure of its identity.

ARTHROPTERUS POLITUS.

Long. 5½ lin. lat. 1¾ lin.

Rufo-piceus nitidus leviter punctatus, antennis angustis articulo primo parvo oblongo, oculis pallidis, capite postice subproducto angulato, thorace quadrato breviter profunde canaliculato ad latera transversim bifoveato, elytris subtiliter punctatis subtruncatis, tibiis angustis.

Hab. Liverpool Plains.

The joints 2 to 9 of the antennæ are in this species little broader than the length, and the tibiæ are as narrow as those of *A. Wilsonii*. It most resembles *A. Kingii*, but differs inter alia in the broader and squarer thorax.

ARTHROPTERUS LATIPENNIS.

Long. 5¼ lin., lat. 2 lin.

Rufo-piceus nitidus latus subplanus leviter punctatus, antennis angustis articulo primo quadrato, capite inter oculos biimpresso, thorace subquadrato canaliculato ad latera antice late rotundato ciliato postice angustato obsolete transversim impresso, elytris latis longis apice late subrotundatis, tibiis angustis extus acute terminatis.

Hab. Flinder's Range, South Australia.

The antennæ in this species are even narrower than the last, while the tibiæ are broader, but its most distinctive character is the large, broadly rounded thorax. It is unique in the Museum.

The genus *Phymatopterus* Westw. contains the species *piceus* and *Macleayi* of Westwood, the latter originally taken by him for *Arthropterus Macleayi* Don. Considerable confusion exists as regards the identity of Donovan's species, and mistakes have, consequently, been frequently made.

As I possess the original specimen from the Cabinet of Francillon, I am enabled to speak with certainty on the subject. *A. Macleayi* is a narrow species, with elongated thorax, and is found about Rope's Creek and the country lying between Parramatta and the Nepean.

I defer the descriptions of one or two new *Phymatopteri* for the present.

The genus *Anoplognathus* MacLeay includes a number of the largest and most showy Australian Coleoptera. As might be supposed, it has therefore received a tolerably large share of attention, and but few species remain to be described. While affixing names to, and writing descriptions of the nevelties in my collection, it has occurred to me that it might be useful to take the opportunity of making a general revision of the group. In doing so, I shall at present limit myself to the section of the family which have the mesosternum produced in front—the *Anoplognathides vrais* of Lacordaire. The Australian species of this group are at present confined to the genera *Anoplognathus* and *Repsimus* of MacLeay.—Hor. Ent. 1, p. 143-144. I propose to add to these the genus *Calloodes* White, which, though never properly described, and not acknowledged by naturalists, may, I think, be very properly used for some intermediate insects, having the broad head and Dytiscus like form of *Repsimus*, but without the enlarged hind legs.

The following synopsis of the first genus *Anoplognathus* is in accordance with Burmeister's plan of subdivision:—

Section 1.—Pygidium for the most part naked in both sexes, sometimes with fine scattered hairs at the sides, or tufted at the apex.

Subsection 1.—Each clytron rounded at the apex, forming a deep emargination at the suture.

    a. Pygidium finely acuducted.

        Anoplognathus viridiæneus Don.

    b. Pygidium smooth.

        Anoplognathus reticulatus Boisd.

            ,,   *rhinastus* Blanch.

            ,,   *longipennis* n. sp.

Subsection 2.—Elytra conjointly rounded at the apex, scarcely emarginate at the suture.

a. Pygidium smooth and strongly bearded at the apex.

Anoplognathus analis Dalm.

,,   *Olivieri* Dalm.

,,   *Duponti* Boisd.

,,   *montanus* n. sp.

,,   *viridicollis* n. sp.

b. Pygidium sparingly bearded.

Anoplognathus rugosus Kirby

,,   *pectoralis* Burm.

,,   *dispar* n. sp.

c. Pygidium acuducted.

Anoplognathus choloropyrus Drapiez.

,,   *Boisduvallii* Boisd.

Section 2.—Pygidium in both sexes, for the most part finely and equally clothed with hair.

Subsection 1.—The mesosternal process long and acute.

Anoplognathus porosus Dalm.

,,   *pallidicollis* Blanch.

,,   *velutinus* Boisd.

,,   *nebulosus* MacLeay, W.

,,   *concolor* Burm.

,,   *rubiginosus* n. sp.

,,   *Odewahnii* n. sp.

,,   *abnormis* n. sp.

Subsection 2.—The mesosternal process short.

a. The process obtuse.

Anoplognathus brunnipennis Gyll.

,,   *flavipennis* Boisd.

b. The process acute.

Anoplognathus hirsutus Burm.

,,   *suturalis* Boisd.

The first of these species *viridiœneus* is a well known Sydney insect. It does not seem to extend far inland, nor is it to be found in Victoria; but I have a small variety, in my cabinet, which is said to have been taken in Tasmania. I have also seen specimens from Wide Bay, Queensland.

A. reticulatus Boisd. I have never to my knowledge seen.

A. rhinastus Blanch. comes from Tasmania. I find two specimens of it in the collection of the late W. S. MacLeay, labelled as variety B of *analis*, a species which it much resembles.

A. longipennis a new species, the description of which with the other new ones, will be given further on, is a Sydney insect.

A. analis Dalm (*viriditarsis* Leach) is an insect of very wide range and considerable variety of colouring.

A. Olivieri Dalm. (*impressus* Boisd.) has also a wide range, and is subject to variety.

A. Duponti Boisd. is evidently identical with an insect in the late Mr. W. S. MacLeay's collection, labelled *A. Olivieri*, variety B, from Van Dieman's Land.

A. montanus nov. sp. I have found at Monaro and Bathurst, and I have a specimen which I believe to be of the same species from Victoria. It has probably a very wide range over the inland parts of the country.

A. viridicollis nov. sp. comes from Darling Downs.

A. rugosus Kirby seems to be found all over the colonies.

A. pectoralis Burm. I have from Monaro and Braidwood; it looks very like one of the many varieties of *A. porosus.*

A. dispar nov. sp. seems to be rather rare. I do not know from what part of New South Wales I procured my specimens.

A. chloropyrus Drapiez. (*nitidulus* Boisd.) is common everywhere here and in Victoria.

A. Boisduvallii Boisd. (*pulchrips* Burm. and *lineatus* MacLeay, W.) is found all over Queensland, and is very subject to variety.

A. porosus Dalm. (*inustus* Kirby) is found everywhere in abundance, and is the most subject to variety of all the species.

A. pallidicollis Blanch. I believe to be merely a variety of *porosus.*

A. velutinus Boisd. is pretty generally distributed, but is not a common insect.

A. nebulosus MacLeay, W., is rare. I have specimens both from Port Denison and Rockhampton.

A. concolor Burm. is from the Clarence River.

A. rubiginosus nov. sp. is from New England.

A. Odewahnii nov. sp. from South Australia.

A. abnormis nov. sp. from Wide Bay.

A. brunnipennis Schonh. I have never taken this insect; the only specimens I have seen are in the collection of the late Mr. W. S. MacLeay, and are labelled "New Holland."

A. flavipennis Boisd. is taken at the Blue Mountains.

A. hirsutus Burm. is a Victorian insect.

A. suturalis Boisd. is common in New South Wales, Victoria, and Tasmania.

ANOPLOGNATHUS LONGIPENNIS.

Long. 11 lin.

Luteus nitidus, capite thorace scutelloque æneo-nitentibus punctulatis, elytris longis pygidium tegentibus rugose punctatis lateribus parallelis apice singulatim rotundatis callo subhumerali brunneo, pedibus æneo-rufis, tarsis viridibus, corpore subtus viridi-nigro nitidissimo albido-piloso mesosterno acute producto, pygidio viridi-aureo marginibus hirtis.

I have only a female specimen of this insect. The parallel sided and singly rounded elytra are its most marked characteristics.

ANOPLOGNATHUS MONTANUS.

Long. 13 lin.

Luteus subæneus nitidus, capite subtiliter punctato clypeo producto reflexo, thorace lævi ad latera leviter punctato, elytris obsolete striatis leviter subrugose punctatis apice rugosis, pedibus rufis, tarsis nigris, corpore subtus nigro albido-piloso, mesosterno elongato subacuto punctato, pygidio rufo ad latera punctato apice barbato.

The clypeus of the male is like that of *A. analis*, and indeed there is such a general resemblance that were it not for some important differences I should have looked upon it as one of the many varieties of that species. The points of difference are the head and clypeus are more finely and the elytra more rugosely punctured, the under surface, which is entirely black, is more densely clothed with white hair, particularly on the sides of the abdominal segments, and the mesosternal process is less acutely pointed and is punctured to the very apex.

Considerable variety as to colour seems to exist in this species, the luteus colour of the thorax passing in some specimens into a brassy green.

ANOPLOGNATHUS VIRIDICOLLIS.

Long. 11½ lin.

Æneo-viridis nitidissimus clypeo testaceo subtiliter punctato brevi late reflexo, elytris pallide testaceis subnitidis rugose productis, pedibus rufis tarsis viridi-nigris, corpore subtus viridi-nigro nitido albido-piloso, mesosterno subrufo elongato acuto lævi, pygidio purpureo-rufo apice barbato lateribus punctatis subacuductis.

The form of the clypeus in the male is very distinct in this species, it is of a testaceous colour, and though not produced so much as in the last species, is largely reflexed in front at a right angle to the base. The head, thorax, and scutellum are of a brilliant golden green, the elytra are of a dull pale testaceous colour and rugosely punctate, and the pygidium is acuducted on the sides.

ANOPLOGNATHUS DISPAR.

Long. 10 lin.

Viridi-æneus nitidus, capite antice dense punctulato, thorace leviter punctulato, elytris flavis rugose punctatis apice singulatim subrotundatis, pedibus fulvis subviridi-æneis tarsis viridi-nigris, corpore subtus viridi-nigro albido-piloso, mesosterno triangulariter elongato subacuto, pygidio viridi-aureo nitidissimo apice subbarbato.

This insect is not unlike a pale variety of *A. rugosus*, it differs in having the elytra less deeply punctured, with the apex of each somewhat rounded. The thorax also is more transverse, and the mesosternal process is less acute and more triangularly pointed. One of this species in the collection of the late Mr. W. S. MacLeay, is labelled *A. porosus* Schonh. which it certainly is not unless indeed the *porosus* of Schonherr is very different from that of Boisduval and other authors, and from the *inustus* of Kirby.

ANOPLOGNATHUS RUBIGINOSUS.

Long. 10 lin.

Breviter ovatus convexus, capite thorace scutelloque æneo-rufis subviridibus nitidissimis leviter punctulatis, maris clypeo late truncato et cum lateribus reflexo, elytris purpureo-rufis dense subseriatim punctatis, pedibus rufis subpurpureis tarsis piceis, corpore subtus nigro nitido flavo-piloso, mesosterno acute producto, pygidio magno punctato triangulari viridi-nigro hirto.

This very distinct species bears no resemblance to any other I know.

The short oval convex form, square clypeus, and large hairy pygidium will serve at once to distinguish it.

ANOPLOGNATHUS ODEWAHNII.

Long. 14 lin.

Testaceo-æneus subviridis nitidus, capite dense punctulato maris clypeo longo reflexo antice dilatato apice rotundato, thorace leviter punctulato medio subtiliter canaliculato ad latera subangulariter rotundato, elytris pallidioribus obsolete striatis dense leviter punctatis ad suturam brunneis apice subrotundatis minime emarginatis angulo suturali rotundato, pedibus rufis tarsis piceis, corpore subtus viridi piloso, mesosterno sublonge producto subacuto piloso, pygidio rufo acuducto pubescente.

The above description is taken from a male, what I believe to be the female is a very different looking insect. It is broad flat subnitid, and of a pale testaceous colour. The mesosternal process is long, acute, and of a golden lustre, as are also the thighs and pygidium, in all else the two sexes are alike.

ANOPLOGNATHUS ABNORMIS.

Long. 7 lin.

Elongato-ovatus pallide testaceus subnitidus, maris clypeo late rotundato anguste reflexo, capite lato plano punctulato maculis duabus viridi-nigris ornato, thorace transverso punctulato quadrivittato—vittis viridi-nigris externis undulatis—basi viridi-marginato, scutello late viridi-marginato, elytris dense subscriatim punctatis, femoribus testaceoæneis tibiis tarsisque rufis, corpore subtus viridi-nigris nitidis parce albido-pilosis, mesosterno acute producto, pygidio dense pubescente.

The broad rounded head of this insect gives it something of the appearance of a *Calloodes*. *A. concolor* Burm. is the species it most resembles.

The genus *Calloodes* White will comprise *Greyanus* White, the species for which this genus was originally proposed, a very rare and beautiful insect from Port Denison, *Rayneri* mihi. also from Port Denison, a species placed by Gemminger and Harold in the genus Repsimus, and *æneus* and *Atkinsonii* of Waterhouse, both from Rockingham Bay. The first of these (*æneus*) I have never seen, but from the description I have little doubt that it has the *Calloodes* form.

To those four species I now add the two following :—

CALLOODES PRASINUS.

Long. 10 lin.

Viridis subopalescens subnitidus, clypeo late rotundato in mare antice et lateraliter reflexo in femina plano punctato, thorace lateraliter punctulato, elytris obsolete striatis leviter punctulatis pedibus rufo-aureis, corpore subtus viridi parce piloso, mesosterno producto subobtuso apice subrecurvo, pygidio viridi punctato subacuducto.

Hab. North Australia.

CALLOODES MASTERSII.

Long. 8 lin.

Argenteo-viridis nitidissimus, capite dense punctulato, clypeo
rotundato in mare angulato, thorace leviter punctulato,
elytris leviter punctulato-striatis, pedibus pallide rufis, corpore subtus parce albido-piloso, mesosterno minime producto
obtuso, pygidio acuducto apice subbarbato.

Hab. Port Denison.

The remaining genus *Repsimus* will by this arrangement
comprise only three species—*æneus* Fab. found all over New
South Wales and Queensland, *manicatus* Swartz. of equally wide
range, and *purpureipes* mihi. from Gayndah, Queensland.

The genus *Bolboceras* Kirby is well represented in Australia.
The species described, however, fall far short of the numbers to
be found in collections.   Mr. Bainbridge, and subsequently
Professor Westwood, described some years ago a number of them,
but almost exclusively from Western Australia.   The species of
South Australia seem to have been quite overlooked; and the
same may be said of those of the North, with the exception of a
few in Mr. Hope's collection from Port Essington, and of three
described by me of late years, from Port Denison and Gayndah.

I shall commence my descriptions of the new species with
those which have two or more prominent horns on the head in
the male.

BOLBOCERAS ARMIGERUM.

Long. 9 lin.

Rufum nitidum, capite antice transversim elevato tricornuto
postice utrinque excavato laevi, thorace antice retuso verti-
cali transversim quadricornuto ad latera rugose punctato
medio subobliterate canaliculato utrinque prope medium
alte foveato, elytris striatis striis subtilissime punctatis,
corpore subtus dense fulvo-hirto, tibiis anticis extus sex
dentibus armatis.

Hab. Rockhampton.

I have only the male of this species.   The head is furnished in
front with a transverse elevation, terminating in three strong

horns, the middle one straight, the others diverging ; in front of
this elevation, and near its base, there is a triangular transverse
"carina," joined at the apex of the triangle by a straight carina
proceeding from the middle horn ; behind this the head is
smooth and concave on each side. The thorax is shortly retuse
and perpendicular in front from side to side, and is armed on the
summit of the retuse portion with four strong pointed horns ;
two near the centre, diverging considerably, and one near each
anterior angle. The sides are roughly punctuate, and behind
and a little outside of each of the centre horns there is a very
deep fovea. The elytra are very finely striated, with the interstices smooth.

BOLBOCERAS PUNCTICOLLE.

Long. 6½ lin.

Piceum nitidum capite antice transversim semicirculariter
elevato apicibus productis postice inter oculos bicornuto,
thorace antice parum retuso pone oculos alte foveato
omnino intervallis punctatis parce instructo, elytris striato-
punctatis tibiis anticis extus 5 dentatis.

Femina capite antice transversim elevato tridentato, postice
inter oculos bituberculato, thorace vix retuso haud alte
foveato.

Hab. South Australia.

The male has a transverse elevation in front, produced at each
side into a strong horn, and behind, between the eyes, two
strong, conical horns. The thorax is a little retuse in front, with
a very deep fovea on each side of the retuse portion, and with
patches of rough punctures in various portions of the surface.
In thefemale, the head is tridentate in front, and bituberculate
behind, and the thorax is without the deep fovea.

BOLBOCERAS LATICORNE.

Long. 6 lin.

Piceo-rufum nitidum capite antice punctato bicornuto—cornubus fortibus subdivergentibus—postice lævi ad latera
utrinque bidentato, thorace brevi punctato antice medio

retuso valde excavato—excavatione postice carina transversa et utrinque lato bifido cornu instructa—elytris striato-
punctatis interstitiis subconvexis subtiliter punctatis, tibiis
anticis extus sex dentatis.

Femina capite bituberculato clypeo apice bidentato, thorace
haud profunde excavato—excavatione antice utrinque tuberculo instructa.

Hab. South Australia.

The head in the male is furnished with two strong divergent
horns, united at the base in a semicircular transverse elevation,
and is bidentated on the sides, the posterior dentation being large
and triangular. The thorax has a very deep excavation in front,
extending nearly to the base, which is bounded behind by a transverse carina, on each side by a broad horn bifid at the apex, and
in front by an extension of the apical border.

The female differs from the male in having the clypeus bidentate, the horns of the head mere tubercles, and the thorax less
excavated, with tubercles in the place of the broad horns of the
other, and the apical margin distinctly tuberculated in the
middle. Seven teeth may be counted on the fore tibiæ of this
species, but the uppermost one is very minute.

The following have only one horn on the head :—

BOLBOCERAS ANGULICORNE.

Long. 10 lin.

Piceo-rufum nitidum capite antice cornuto—cornu longo subrecurvo postice medio triangulariter dentato antice ad basin
carina sinuata instructo, thorace brevi valde retuso ante
medium bicornuto—cornubus latis subtruncatis subrecurvis
—ad latera punctato subfoveato, elytris striato-punctatis,
tibiis anticis extus 5 dentatis.   .

Hab. Port Curtis.

Though differing in many ways there is a close affinity in this
insect to *B. Rhinoceros* mihi. The chief differences are the
angular tooth on the back of the horn of the head, the much
bisinuated "carina" at the base of the horn in front, the very
broad thoracic horns, and the more deeply striato-punctate
elytra. I have not seen the female.

BOLBOCERAS CAVICOLLE.

Long. 8 lin.

Piceo-rufum nitidum, capite antice cornuto—cornu longo subrecurvo postice medio triangulariter dentato, thorace antice
medio retuso longitudinaliter profunde excavato utrinque
breviter cornuto ad latera rugose punctato, elytris leviter
striato-punctatis, tibiis anticis extus 6 dentatis.

Femina cornu capitis brevi haud postice angulato, thorace
leviter retuso bituberculato.

Hab. South Australia.

Like the last, the horn on the head is long, slightly recurved,
and triangularly toothed in the middle of the posterior surface.
The thorax has a long deep excavation in the middle in front
which extends nearly to the base, and has a short subrecurved
horn on each side of the excavation. In the female the horn of
the head is shorter and without the angular tooth behind, and the
thorax is only slightly retuse and excavated, with tubercular
prominences in place of the horns of the male.

BOLBOCERAS CORNIGERUM.

Long. 9 lin.

Piceo-rufum nitidum capite antice cornuto—cornu elongato
subacuto antice basi bisinuatim carinato, thorace antice
medio retuso modice excavato—excavatione lævi utrinque
cornu brevi instructa—ad latera rugose punctato foveolato,
elytris striato-punctatis, tibiis anticis extus 5 dentatis.

Femina capite cornu brevi bifido instructo, thorace antice verti-
cali lævi.

Hab. Swan River.

The horn of the head is long, pointed, and without any tooth
on the posterior surface. The transverse carina at the base of
the horn in front is emarginate in the middle, and much sinuated
at each side, as in *B. angulicorne*. The thorax is only moderately retuse and concave in front in the middle, and has a short,
obtuse horn on each side of the retuse portion. The sides are

very coarsely punctured, and there is a small deep fovea near each posterior angle. The female has no excavation in the thorax, and the vertex is furnished with a short bifid horn.

BOLBOCERAS CARPENTARIÆ.

Long. 8½ lin.

Piceo-rufum nitidum, capite antice cornuto—cornu subbrevi subacuto antice basi bisinuatim carinato, thorace vix retuso bicornuto punctato ad latera foveolato basi lævi, elytris striato-punctatis, tibiis anticis extus 5 dentatis.

Hab. Sweer's Island.

I have only seen the male of this species. The horn on the head is shorter than in the last. The thorax is scarcely excavated but has a plane surface in front, with a short somewhat porrect horn on each side. The base, with the exception of the median line, is smooth, the remainder rugosely punctate and deeply foveated on each side.

BOLBOCERAS DENTICOLLE.

Long. 9½ lin.

Piceo-rufum nitidum, capite punctato apice cornuto ad latera utrinque bidentato, thorace subretuso apice subobtuse tuberculato-producto medio fortiter bicornuto—cornubus divergentibus—omnino parce punctato lateribus serratis, elytris fortiter striato-punctatis, tibiis anticis extus 6 dentatis.

Hab. Victoria River, or Peak Downs.

The male of this insect has a short nearly upright horn at the very extremity of the head, with two teeth on each side, one large and triangular near the eyes, the other smaller between that and the clypeus. The thorax has an obtusely pointed tubercle at the apex with a somewhat excavated space behind, which is surmounted by two strong divergent horns, the sides are dentated. The elytra are strongly striato-punctate. I have not seen the female. This species is in the Museum collection, and was brought from one of the above named localities by the Expedition which, under the command of the late Sir Thomas Mitchell, first penetrated to the Victoria river.

BOLBOCERAS LACUNOSUM.

Long. 8 lin.

Piceo-rufum nitidum, capite punctato antice transversim carinato obsolete tridentato, thorace punctato subplano haud
retuso medio subtriangulariter excavato excavatione antice
postice et utrinque tuberculo instructa, elytris nigris fortiter
striato-punctatis interstitiis subconvexis, tibiis anticis extus
sexdentatis.

The peculiarity in this species is the horizontally placed thorax
with a large triangular or cordiform excavation in the middle, at
the apex of which there is a small horn, a tubercle on each side,
and a rounded prominence behind. It is found near Sydney.
This and the two following insects are without horns on the head.

BOLBOCERAS PLANICEPS.

Long. 4 lin.

Rufum nitidum capite plano lævi antice punctato triangulariter transversim carinato tenuiter tridentato, thoraco
lævi minime retuso margine apicali bifoveato linea media
antice impressa ad latera antice rugose punctato medio sub-
transversim punctato-foveolato, elytris 8 vel 9 striatis
striis leviter punctatis interstitiis planis, tibiis anticis extus
sexdentatis dentibus duobus basalibus subobsoletis.

Hab. Sweer's Island.

The flat, smooth head, and the limited number of striæ on the
elytra will serve at once to enable any one to recognize this
insect. The median line is broadly marked, on the anterior and
slightly retuse portion of the thorax, there are two small foveæ
on the anterior edge, and an elongate punctured one near the
middle of each side.

BOLBOCERAS SWEERII.

Long. 3½ lin.

Piceum nitidum, capite punctato medio obtuse tuberculato
antice subtriangulariter leviter carinato, thorace parce

punctato antice minime retuso medio leviter canaliculato
ad latera utrinque bifoveato, elytris leviter striato-punctatis, tibiis anticis extus sex-dentatis.

Hab. Sweer's Island.

This differs from the last species in having the head punctured,
in having a tubercle on the forehead, in having the thorax thinly
punctured, not retuse, and lightly canaliculate in the middle, and
in having the usual number of striæ on the elytra.

In a previous portion of this paper I described a number of
new species of *Promecoderus* and allied genera. Since then, I
have received from Peak Downs a very extraordinary looking
insect, evidently one of the *Cnemacanthidæ*, but very distinct
from any of the genera of that group. From the massive character
of the prosternum I propose for it the generic name

BRITHYSTERNUM.

Caput magnum longum subnutans.

Mentum profunde emarginatum, medio edentato, lobis lateralibus magnis extrorsum rotundatis introrsum erectis.

Labium antice subrotundatum subtilissime emarginatum bisetosum paraglossis adhærentibus haud longioribus.

Maxillæ acutæ arcuatæ.

Palpi maxillares subcrassi articulo penultimo brevi, ultimo
truncato.

Palpi labiales sublongi crassi truncati.

Labrum magnum transversum truncatum.

Mandibalæ haud magnæ subarcuatæ.

Antennæ subsetiformes articulo primo crasso, tertio ceteris
longiori.

Thorax elongatus.

Elytra subconvexa angusta.

Pedes graciles, femoribus anticis longis subarcuatis : intermediis
subtus versus apicem calcaratis : posticis trochanteribus
longis acuminatis, tibiis anticis arcuatis, tarsis anticis in
mare triangularibus.

Prosternum latum retrorsum porrectum.

Corpus longum subcylindricum pedunculatum apterum.

The description given above will show that in some respects this genus resembles *Parroa* of Castelnau. The very elongate form, slight legs, and largely produced sternum gives it, however, a very different aspect.

BRITHYSTERNUM CALCARATUM.

Long. 12 lin., lat. 3 lin.

Nigrum nitidum lœve, capite subconvexo inter oculos leviter bi-impresso, clypeo utriuque profunde bipunctato, thorace longo antice truncato capite latiori angulis rotundato-productis, medio canaliculato, postice truncato subangustato transversim impresso lateribus marginatis vix rotundatis, elytris haud thorace latioribus punctis paucis marginalibus, tibiis anticis arcuatis prope apicem emarginatis dente interno apicali magno, femoribus intermediis calcare magno subtus versus apicem armatis, prosterno lato plano retrorsum porrecto subrecurvo.

The antennæ palpi and tarsi are piceous, the rest of the insect is black, nitid, and smooth. The head is a little convex, has two large shallow foveæ between the eyes, and two small but deep punctures on each side of the clypeus. The labrum is canaliculate in the middle, and has its anterior edge marked with strong setigerous punctures. The mandibles are lightly striolate longitudinally, and the right one is furnished with a tooth in the middle. The eyes are round and rather prominent, and there is a small deep puncture behind and a little above each eye. The antennæ are much shorter than the head and thorax united, the first joint is thick, the remainder are all obconical, the second joint being the smallest, and the third a little the largest, and gradually tapering to the apical joint which is acuminate. The thorax is slightly broader than the head in front, nearly twice as long as wide, truncate in front and behind, margined at the sides, obtusely produced at the anterior angles, very slightly rounded on the sides, very little narrowed towards the base, transversely impressed near the base and finely canaliculate on the median line. The elytra are elongate, not wider than the thorax, and smooth with the exception of three or four punctures in the humeral

x

margin. The legs are slender and rather long. The fore thighs and tibiæ are a little curved inwards, the latter is not enlarged towards the apex, has the emargination on the inner surface small and very near the apex, and has the terminal spine on the inner side large, acute and curved. The intermediate thighs are armed on the under side not far from the apex with a large strong slightly curved spur. The trochanters of the hind legs are large, straight, and bluntly acuminate. The four first joints of the fore tarsi in the male are rather broader than in the female. The prosternum is flat and broad, and is extended backwards in a broad laminated and somewhat recurved form, as if by its contact with the mesosternum it was intended to prevent the long heavy head and thorax of the insect from drooping too much.

I received, a few months ago, from the Richmond River district, a *Feronia* of sculpture of a most peculiar and unusual kind. It differs in one way or another from all the subgenera given by Baron de Chaudoir in his " Essai sur les Feroniens de l'Australia," and it does not quite fit into any of the very many subdivisions of the group. *Abax* seems to be the one to which it most approximates, I shall accordingly call the insect

ABAX SULCIPENNIS.

Long. 9½ lin.

Brunneo-cupreus nitidus subplanus, capite profunde bisulcato, thorace subtransverso medio profunde canaliculato postice profunde utrinque longitudinaliter bifoveato—fovea externa lata brevi—ad latera marginato subrotundato basi subtruncato medio minime emarginato angulis anticis rotundatis subproductis posticis rectis, elytris thorace parum latioribus brevibus planis late bisulcatis, sulcis carinatis apice fortiter sinuato lateribus seriatim punctatis, corpore subtus nigro nitido, antennis palpis tarsisque piceis.

If the work so ably commenced by Baron de Chaudoir of arranging and classifying the Australian *Feronidæ*, should ever be brought to a completion, this insect will, I have no doubt, form a distinct subgenus. It differs from the most common group of the larger Australian *Feronic*, named by Chaudoir *Notonomus*, in having the third joint of the antennæ much longer than the fourth, from *Rhytisternus* of the same author in the thicker and more truncate palpi, while from *Homalosoma*, *Trichosternus*, *Prionophorus*, *Loxodactylus*, and *Chlænioidius* of Chaudoir, it is still wider apart.

The sculpture of the elytra is unlike that of any of the family I have seen, more resembling, in fact, that of a true *Carabus* than of a *Feronia*.

Each elytron has two broad, deep grooves, extending from the base to the apex, and in the bottom of each groove there is a fine rounded "carina." Near each lateral margin there is a row of large more or less distant punctures. The extreme apex is rounded, but the sides a little above the apex are profoundly sinuate.

My only specimen of this insect is a female. It was found, I believe, in the upper valley of the Richmond River.

––––––

When on an Entomological excursion to the Murrumbidgee in the spring of last year, Mr. Masters and I captured, in considerable numbers, in the neighbourhood of Mundarlo and Tarcuttah, a species of *Tmesiphorus*, which we invariably found in the society of a small red ant. So invariable was the association that whenever on turning over a log we found some of the ants we knew that a search in their passages would certainly lead to the discovery of some of these attendant beetles.

The ant answers very nearly to the genus Ectatomma of F. Smith. It is undescribed, I give it therefore a name and description.

ECTATOMMA SOCIALIS.

Long. 2½ lin.

Piceo-rufa subtilissime dense punctulata parce aureo-pubescens, capite quadrato subconvexo subtus hirto, oculis parvis lateralibus ante medium positis, abdominis petiola brevi lata postice truncata, segmentis terminalibus subhirtis, pedibus longis flavis.

I have only seen the worker of this species. It seems to differ somewhat in the size and position of the eye from F. Smith's description of the genus.

For the attendant beetle I propose the name of

TMESIPHORUS FORMICINUS.

Long. 1¼ lin.

Piceo-rufus dense punctulatus aureo-pubescens, capite convexo bipunctato, thorace convexo rotundato, elytris convexis leviter bisulcatis, abdomine marginato postice bicarinato.

This species differs from *T. Kingii* mihi, in being of a lighter colour, finer puncturation, less deep sculpture, denser pubescence less elongate thorax, and in having the 9th and 10th joints of the antennæ of equal size. From the only other Australian species of the genus *T. Macleayi* King. it differs in being of a darker colour denser puncturation, denser pubescence, and in having the thorax more round and less elongate.

INDEX.

INDEX.